AF700286

SOIL AND ROOT DAMAGE IN FORESTRY

SOIL AND ROOT DAMAGE IN FORESTRY

Reducing the Impact of Forest Mechanization

IWAN WÄSTERLUND
Olasgarden Roads and Forestry, Sävar, Sweden

ELSEVIER

Elsevier
Radarweg 29, PO Box 211, 1000 AE Amsterdam, Netherlands
The Boulevard, Langford Lane, Kidlington, Oxford OX5 1GB, United Kingdom
50 Hampshire Street, 5th Floor, Cambridge, MA 02139, United States

Copyright © 2020 Elsevier Inc. All rights reserved.

No part of this publication may be reproduced or transmitted in any form or by any means, electronic or mechanical, including photocopying, recording, or any information storage and retrieval system, without permission in writing from the publisher. Details on how to seek permission, further information about the Publisher's permissions policies and our arrangements with organizations such as the Copyright Clearance Center and the Copyright Licensing Agency, can be found at our website: www.elsevier.com/permissions.

This book and the individual contributions contained in it are protected under copyright by the Publisher (other than as may be noted herein).

Notices

Knowledge and best practice in this field are constantly changing. As new research and experience broaden our understanding, changes in research methods, professional practices, or medical treatment may become necessary.

Practitioners and researchers must always rely on their own experience and knowledge in evaluating and using any information, methods, compounds, or experiments described herein. In using such information or methods they should be mindful of their own safety and the safety of others, including parties for whom they have a professional responsibility.

To the fullest extent of the law, neither the Publisher nor the authors, contributors, or editors, assume any liability for any injury and/or damage to persons or property as a matter of products liability, negligence or otherwise, or from any use or operation of any methods, products, instructions, or ideas contained in the material herein.

Library of Congress Cataloging-in-Publication Data
A catalog record for this book is available from the Library of Congress

British Library Cataloguing-in-Publication Data
A catalogue record for this book is available from the British Library

ISBN: 978-0-12-822070-2

For information on all Elsevier publications visit our website at https://www.elsevier.com/books-and-journals

Publisher: Candice Janco
Editorial Project Manager: Alex Ford
Production Project Manager: Bharatwaj Varatharajan
Cover Designer: Miles Hitchen

Typeset by TNQ Technologies

Working together to grow libraries in developing countries
www.elsevier.com • www.bookaid.org

To my wife Dianne and my student Dr. Linnea Hansson

Contents

Preface

This book is made as a summary of my research and work during a 30-year period working and experimenting with reduction of root and soil damages in forestry landscape being one of the experts in Europe on soil damage at FAO/ECE/ILO and IUFRO P3. The first two chapters were printed at a seminar and workshop as a seminar in Feldafing, Germany held by Bundesministerium für Ernärung, Landwirtschaft und Forsten as a mimeograph.

Below is an example of mainly students taking a path to the bus stop with both wearing of vegetation and some soil compaction. Like Lull (1959) described 60 years ago, even cattle may compact the soil and the same thing happened here also.

The fact that humans by simply walking on a path or making a new path just to get to the buss changed the vegetation and compacted the soil (Photo Iwan Wästerlund).

A Czeckish experiment where soil was just blown away to a.10 cm depth and root damages monitored after passage with a farm tractor with a trailer loaded with timber. Monitoring was done by following the flow of water transport along the roots up into the stem and showed that a simple root damages impeded water transport on that side of the stem. This experiment was done by Drs Ulrich and Cermak (Photo Iwan Wästerlund).

The wheel rut marked with strings on right side.

This compilation of literature and own studies that have been collected during 50 years. It started when foresters questioned me during 1980s and I collected a group of specialists to FAO from around the world, but it turned out that we could not agree on the drawings. Thus, it was just for me to give up, but since I both conducted number of experiments and collected a huge number of other studies, so why not put all data together in a book, and here it is.

Because nowadays, it is very popular to talk about it and direct research in that area.

Green people tend to think that we should make a nature reserve of this nice forestry area, and when they come back after 10 years, it looks completely different because the nature is dynamic and ingrowth of trees that has been seeded in and storms can upheave bigger trees and only young trees may be left or a fire could change the whole appearance of the forest.

CHAPTER 1

Forest responses to soil disturbance due to machine traffic

Introduction

In the ISTVS glossary (Anon, 1968), the terrain trafficability is defined as "the ability of terrain to support the passage of vehicles." The idea behind that definition is that a vehicle should be able to pass the terrain with only an acceptable degree of soil disturbance. A such statement will immediately raise the question: What is an acceptable degree of soil disturbance? To say that there should be no disturbance at all is to overdo it. In that case the one who is proposing this must realize that no man or animal should be allowed to walk on the ground (cf. Fig. 1.1)!

From another point of view, we could try to interpret acceptable disturbance as the forest should have a sustainable growth. Sustainability is a very fashionable word, and it is defined by the Brundtland commission as a development that "meets the needs of the present without compromising the ability of future generations to meet their own needs" (WCED, 1987). The question is what does it mean? If the regeneration in terms of species mix, soil conditions, and so on matches the harvested forest area, it is, theoretically, sustainable utilization (Freezailah, 1994). Should the forest look like and grow like it was doing yesterday or 500 years ago? How should we regard a natural development of a forest with time, is that acceptable? Now, with all these questions in mind, we realize that we are back on the starting point. The only useable definition on an acceptable degree of soil disturbance so far is if the site productivity (growth of the stand) is equal after the soil disturbance from machine traffic as it was before. If this definition can be agreed upon, then we have a measure of the

Soil and Root Damage in Forestry
ISBN 978-0-12-822070-2
https://doi.org/10.1016/B978-0-12-822070-2.00001-8
© 2020 Elsevier Inc.
All rights reserved.

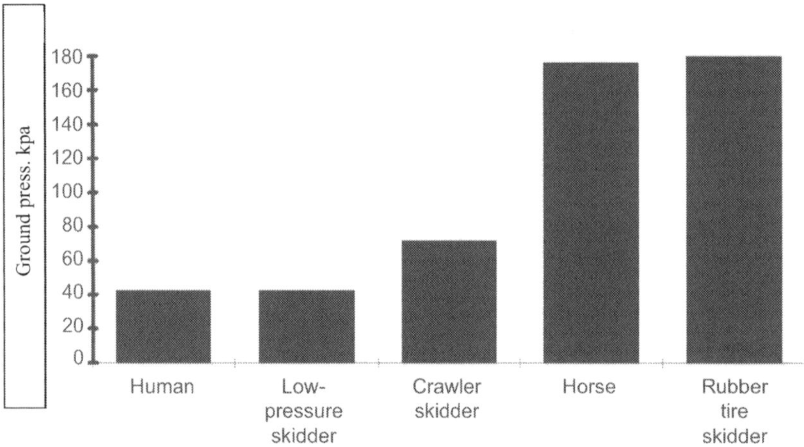

Figure 1.1 Approximate standing ground pressure of a person, horse, crawler, and skidder. *(After Adams, P.W., Froehlich, H.A., 1981. Compaction of Forest Soils. Oregon State University, Extension Service, PNW-217.)*

soil disturbance that can be relatively easily measured and defined, as well as quantified in terms of volume and money. The presentation below takes that approach as a basis and if the reader comes up with better frames to judge from, the reader is welcome to present it. However, as discussed in the section below, there are many other aspects of soil disturbance but they are at the present state of knowledge quite difficult to quantify and may therefore not be a proper justification of measures to minimize soil disturbance at forest operations.

Clearfelling operations

Trafficated area

At clearfellings of forest areas with the tree-length method, the skid trails are reported to take up 15%–35% of the site area but the total area influenced may run as high as 80% (Lull, 1959; Froehlich, 1978; Martin, 1988; Reisinger et al., 1988; Davis, 1992). More than half of the skid trail area (50%–75%) has been classified as disturbed and compacted. By using designated skid trails and increased spacings to 125 ft (37 m), the disturbed area could be limited from 15%–20% to 8% according to Stewart et al. (1988). In that case, a grapple skidder cannot be used. In a controlled logging operation in Malaysia, trails after skidders covered 24% of the area, whereas trails after manual extraction covered only 4% (Malmer and Grip, 1990). One interesting thing is that very few seem to bother about the feller-bunchers or harvesters. However, in a study by Lee et al. (1990), it was found that a frame-mounted feller-buncher caused more soil disturbance and compaction than a boom-mounted because the first machine type has to travel to each tree. Martin (1988) recommends the use of forwarders rather than dragging the whole tree to reduce exposure of mineral soil. He also recommends delimbing of conifers on the site and placing the slash in the trails to reduce compaction.

In Europe, the traffic during clearfelling appears to be a forgotten topic. Roughly estimated, about 30% of the area is used for forwarding or skidding of the timber. With use of a feller-buncher or a harvester the influenced area could be much higher especially if the machines are traveling different routes.

Generally, the clearfelling is followed by some kind of site preparation before planting which may further increase the traffic on the site. The site preparation for the next tree generation could imply slash removal, soil scarification, etc. A heavy work well suited for mechanization which means machines traveling back and forth over the area. Slash piling is often included in the total effects of the logging operation, but Davis (1992) argues that alternatives to tractor slash piling should be considered to reduce the trafficated area. Gent and Morris (1986) concluded that windrowing and chopping of slash added only little added effect to the total soil bulk density partly because of all traffic before that treatment. The organic matter content after slashing may be in the same (low) level as in the skid trails (Snider and Miller, 1985).

However, if the soil preparation is done with, for example, a powered two-row disc trencher or a similar device, the soil is loosened behind the machine. There appears to be no remaining soil compaction due to the machine wheels and previously compacted soil is loosened at least near the surface (Froehlich and McNabb, 1984; Gyldberg, 1993). Martin (1988) recommends "Scarification should be planned carefully and not be a by-product of an otherwise haphazard skidding arrangement."

Effects on the soil

In most reported cases, 50%—100% of the skid trails show an apparent rut formation, a markedly increased bulk density and severely reduced macroporosity as well as water infiltration rates (see, e.g., Campbell et al., 1973; Jakobsen, 1983; Hildebrand and Wiebel, 1986; Incerti et al., 1987; Shetron et al., 1988; Hofmann, 1989; Cullen et al., 1991). The use of crawler tractors are reported to cause more soil disturbance than rubber-tired machines (Clayton, 1990), and although they might have much less nominal ground pressure than the wheeled machines, they might compact the soil to the same degree (Kreh et al., 1985; Clayton, 1990). The reverse is, however, also reported (Shishiuchi and Adachi, 1982). Very few studies are done on soil disturbance after forwarding in clearfelling operations (e.g., Eliasson and Wästerlund, 2007). The soil moisture content during the operation is one of the more important factors for the occurrence of soil disturbance and in moist soils the bulk density may raise up to 50% (Hassan, 1978).

To reduce the severe soil disturbances after logging operations, it is argued that machines should run mainly on designated skid trails so only a minor part of the area is heavily compacted (Adams and Froehlich, 1981). The argument is based on that already the first trip may cause 50%–75% of the compaction achieved after maybe 25 trips as shown in many studies. However, the compaction level may depend on the harvesting system and the machinery used. With use of light and low ground pressure wheeled machines, even a loosening of the soil has been reported during the first two passes (Lenhard, 1986), whereas many passes caused heavy soil compaction. In that case, it may be an advantage to avoid running in the same wheel ruts!

Persistence of soil compaction seems to be very long in North America. Although recovery within 5 years is reported for less severe disturbed areas (Reisinger et al., 1992), it may take up to or more than 30 years in other cases (Wert and Thomas, 1981; Froehlich et al., 1985; Corns, 1988). In Australia, compaction was still evident after 32 years, while in Russia the dry soils recovered after 5–7 years, whereas soils with a rather high water table may remain compacted for more than 15 years (Jakobsen, 1983 and Ivanov, 1976, respectively).

Growth responses

Measurements of tree growth on compacted soils show that significant impacts can and do occur. Seedling height growth has been studied most, with reported growth reductions on compacted soils from many different soils types in the United States ranging from about 5% to 60% (Fig. 1.2). In studies of residual tree growth in thinned stands in the Pacific Northwest, overall volume growth response with compacted soil was reduced by 5%–15%. Individual trees, however, showed growth impacts that have been observed for commercial-sized trees in unthinned stands established on compacted soil (Adams and Froehlich, 1981). Soil disturbance and compaction will affect both diameter and height growth which in some

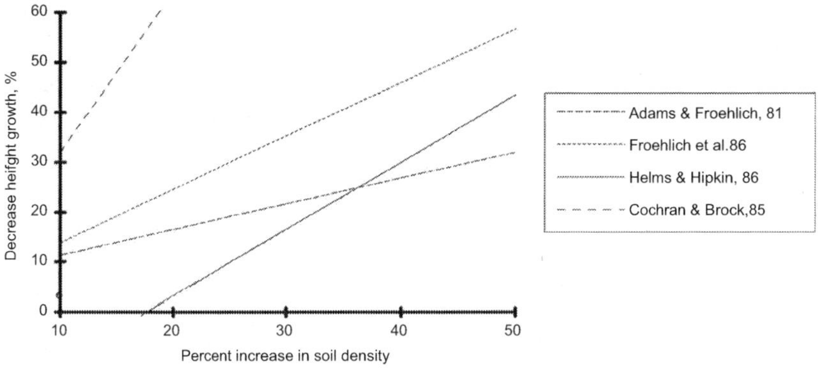

Figure 1.2 Relationship between the increase in forest soil bulk density and the variety of species; growth of tree seedlings. Adams and Froehlich (1981): Seedling height growth, Cochran and Brock (1985): 6-year-old ponderosa pine, height growth; Froehlich et al. (1986): 9- to 16-year-old ponderosa pine, volume growth; Helms and Hipkin (1986): 16-year-old ponderosa pine, volume growth.

cases will result in a further decrease in volume growth (cf. e.g., Froehlich et al., 1986) or in other cases rather similar degrees of reductions in volume as with height growth (cf. e.g., Helms andHipkin, 1986). The growth effects are in many cases rather well correlated to the increase in soil bulk density. However, a direct correlation with the soil bulk density (height growth = 5.7−1.74BD, Clayton et al., 1987, BD = bulk density) is less useful since it would imply that an extremely loose soil would be very good. This is not likely as explained in my later chapter.

Growth reductions due to soil disturbance have been obtained in field experiments for some tree species, like *Pinus ponderosa*, *Pinus taeda*, and *Pseudotsuga menziesii*. In one report, a growth reduction has been observed also for Pinuscontorta (Clayton et al., 1987), but in another report no reductions were found (Froehlich et al., 1986; Wästerlund, 1990). The latter case seems more reliable considering the low susceptibility to soil compaction this species has shown in laboratory studies (see further in later chapter). In Canadian studies, soil disturbance has also shown to reduce growth of *Picea engelmannii, P. menziesii,* and *Abies lasiocarpa* on preconstructed skid roads on slopes but in some cases the soil disturbances lead to increased growth (Smith and Wass, 1979). In Tasmania, no clear growth reductions due to soil compaction were obtained for Eucalypts or Pinusradiata (Williamson, 1990).

It should be kept in mind that in the reports from the United States, we are not only looking at the harvesting equipment but also to site clearing methods, which commonly are not paid any greater attention to. Slash piling is often done after the clearfelling with a bulldozer and then the slash is burned. This means that the organic layer on the soil is reduced by two different operations and the soil fertility might be reduced due to the loss of organic material. Considering only the growth of the trees, the losses of nutrients may be compensated by fertilization (cf. Hatchell, 1981).

In a 32-year-old naturally seeded Douglas-fir stand, Wert and Thomas (1981) measured that the skid roads produced 74% less volume and the transition zone 25% less volume than undisturbed areas. These reductions resulted in an overall loss of volume of 11.8% when applied over the entire area. The reduced productivity on skid trails may reduce the total volume yield 6%–16% maybe for the whole second rotation and result in a reduction of the site productivity with one site class (Froehlich and McNabb, 1984; Stewart et al., 1988). The reduced growth implies that an increase of 100–200 USD/ha in yarding costs could be justified for a yarding system that considerably reduces damage to the soil (Stewart et al., 1988). The amount is too small to justify skyline yarding or aerial systems but enough to promote considerable improvement of existing yarding systems.

In Brazil, with fast-growing Eucalyptus it was very visible that soil compaction in wheel ruts decreased the height growth by a meter, whereas the surrounding trees were more than 2–3 m high 2 years after plantation (own observation).

Thinning operations

Trafficated area and effects on soil and roots

Commercial thinning operations are performed 1–5 times during the rotation of the stand. Usually, temporary roads—strip roads—taken up at first thinning are used also during following operations. The spacings could be from 15 to 30 m. With 2.5–4 m wide strip roads, this would imply a disturbed area of 12%–25% (cf. Eriksson, 1981; Fröding, 1982; Murphy, 1982; Wronski, 1984; Shetron et al., 1988). At row thinning, the disturbed area could amount to 37% (Karr et al., 1987).

Soil damage assessment along strip roads are less frequently done but some reports about rut formation and soil compaction after skidding do exist (e.g.,Wronski, 1984; Benthaus and Matthies, 1993). The change from the tree-length harvesting system to the cut-to-length system and using forwarders for the yarding were reported to reduce the soil damage (McNeel and Ballard, 1992; see also Wronski, 1984). Soil damage due to forwarder traffic is mainly studied as rut formation and deep ruts (>10 cm) has been observed on 5%–30% of the strip roads (Eriksson, 1981; Fröding, 1982). Also here, the harvester is seldom studied separately but ongoing experiments in Germany and Sweden will give some further information about harvesters influence on the soil. At wider spacings (>15–20 m), a harvester with a boom-mounted processor will have problems to reach the trees in the middle between the strip roads. A smaller harvester or a feller could then be used taking the trees in the middle via access roads and put them on a reachable distance for the harvester or a forwarder. Recent field studies indicate that such small harvesters (5–6 tonnes) do very little soil damage on the access roads (Wilpert and Parbs, 1992). To use a crawler tractor with a felling shear, which has to travel to each tree, could increase the disturbed area up to 35% (Karr et al., 1987).

Generally, the bigger the machines are, the more soil disturbance they cause (Myhrman, 1990), but some improvements on later versions may nivellate that statement somewhat. By putting logging residuals from coniferous trees in the wheel tracks, the soil damage can be reduced considerably (Fröding, 1992; Eliasson and Wästerlund, 2007), and a 25–30 cm thick layer can reduce the damage almost to nil (Hildebrand and Wiebel, 1986; Schäffer et al., 1991; Schäfer and Sohns, 1993). This can easily be done when working with a one-grip harvester and cut-to-length methods.

It seems to very few studies of soil recovery after thinning operations. In a German study, the recovery was found to take more than 6 years (Benthaus and Matthies, 1993). However, the recovery was faster on minor compacted soil trafficated by wide tyres (80 cm) than on more compacted soil trafficated by normal (40 cm wide) tyres. Kardell (1978) found recovery of soil compaction in a south Swedish experiment to be between 5 and 10 years.

Wounding

Injuries to the tree that damage the inner bark or deeper layers start a wound reaction (Biggs et al., 1984). Deeper injuries also cause a dry zone formation and the initiation of a compartmentalization process, that is, for example, vessel blocking and barrier formation in the sapwood (Coutts, 1977; Shigo, 1985). This means that a part of the defense mechanisms against invading pathogens depends on the tree species.

Although bark wounds may be attacked by, e.g., canker fungi, it generally requires a removal of the bark to really increase the risk for rot infections. There are many studies showing an increase of rot infections with increased wound size. The infection risk also depends on the season, the locality, and the tree species. Birch, Norway spruce, and Sitka spruce belong to those species that are easily infected. According to Isomäki and Kallio (1974) and Meng (1978), a 100-cm^2 bark peel off on Norway spruce may increase the risk of 40% for rot infection which then spreads upward in the stem at a rate of about 20 cm per year. Deeper injuries that also include wounds or crushing of the wood fibers in the wood cylinder increase the risk of rot infection considerably. The greatest risk of rot infection is in the root collar region. A 200-cm^2 wound which include a damaged wood cylinder in the root collar region has an almost 100% chance of getting infected.

Larch, oak, and pine trees are seldom infected by wood-rotting fungi. However, the wound will reduce the wood quality and the wound reaction blocks the part of the transport paths and thereby also the growth. A study on Scots pine showed that the growth reduction is proportional to the wounded portion of the stem girth (Andersson, 1987). A wounded portion of 1/8−1/4 of the stem girth reduces growth by 13% for at least 10 years. The loss in growth loss becomes more if the wound is located in the root collar region.

Similar results have also been reported for Norway spruce (Isomäki and Kallio, 1974).

The economic effects of stem wounds appear to amount to about 10%−15% loss in the value of the tree (Meng, 1978; Fröding, 1986). Although the calculations are uncertain, they show that there are economical reasons to be careful and to avoid these damages. A rather expensive way is to treat the tree wounds with chemicals (Bonneman, 1979). According to Dimitri (1983), the costs of proper chemicals and the treatment would be a profitable investment (0.8−0.9 DM per tree = 1−1.2 USD).

Roots

The tree roots can be divided into two groups: (i) larger than 20 mm (long laterals) and (ii) less than 20 mm in diameter (very large, and large to fine roots according to Böhm, 1979). Typical of a root system in a thinning stand is some eccentricity but usually four first-order long lateral roots have a diameter of more than 20 mm at a distance of 1 m from the trunk base (Björkhem et al., 1975).

The tensile strength of roots seems to be about 2–5 kN cm^{-2} cross-sectional area (Turmanina, 1965; O'Loughlin, 1974; Burroughs and Thomas, 1977; O'Loughlin and Watson, 1979,1981). Thus, the very large roots do generally not break beneath a machine wheel but the bark may be peeled-off.

The strength of the root bark varies with season. During the sap period, the shear strength is 40 N cm^{-2} for Norway spruce and Scots pine, but in the autumn, it increases to 60–80 N cm^{-2} (Wästerlund, 1986). If the wheel contact area is 40 cm^{-2}, maximum tractive forces should be less than 1.6–3.2 kN to avoid bark peel-off (cf. > 6 kN in tensile strength for a root 20 mm in diameter). With 1 kN propelling force with a medium forwarder on level ground, it means 25 N per cm^2 (Olsen and Wästerlund, 1989, 1 kN torque per wheel), it means the root may hold the for the load but bark may be ripped off in summer time.

According to Nilsson and Hyppel (1968), wounds occurring on roots larger than 20 mm in diameter and less than 0.7 m from the trunk base may result in a rot infection that enters the stem. The probabilities for rot infections, rate of dispersal, etc., are not so well established, but they appear to be similar to that of for stem wounds. If long lateral roots are broken, the growth reduction of the tree could be 50% and the growth rate of the rot infection could be twice the rate of the stem infection (Isomäki and Kallio, 1974). However, the effect of root wounds on tree growth may be over-estimated, since those trees are most likely also subjected to soil compaction.

The fine and medium-sized roots may be crushed or torn off beneath a driven wheel (cf. Wronski, 1984). The total number of tree roots may be at least 60–120 roots m^2 and the total length may be 200–500 m m^2 (Wästerlund, 1989, Fig. 1.3). Less than 7% of them are larger than 10 mm in diameter. Damage to these roots is not likely to cause stem rots but may reduce the tree growth.

The stem damage surveys in Sweden in the beginning of the 80-ties was not encouraging (Table 1.1) and comparable with the 14%–18% stem damage measured in a Canadian study (McLaughlin and Pulkki, 1992). The forest companies started an education program to learn and motivate the operators to avoid stem damage, selected machinery more suited for thinning operations and some companies gave also a salary bonus if a certain damage level was not exceeded. As seen in Table 1.1, the operation succeeded.

Figure 1.3 Spruce roots in a thinning stand 10 cm down in the humus layer.

Table 1.1 Frequency of stem damage in different thinning methods (Fröding, 1992).

Method	Damage frequency, % 1980–82	D:o % 1988
Motor manual	8.0	3.0
Processor	14.5	7.5
One-grip harvester	—	5.9

Growth responses

McCreary and Perry (1983) showed that the positive growth responses (edge effects) in strip thinning extended about 3 m into the stand. The extension of the edge effects depends also on the tree size or age (cf. Fig. 1.4). Similarly, the extension of growth reduction due to soil and root damage in a stand should depend on how much of the tree root system that is influenced but now negatively. To be able to compare the growth effects and its extension in stands of different age, a root growth model (Fig. 1.4) was developed for the average effective horizontal spread of the root system (Wästerlund, 1983,1987). The model may represent medium or poorer site conditions in the Nordic countries, but seems to overestimate the root spread on rich sites.

The closer the tree is standing near to the wheel track, the bigger part of the root system may be disturbed, and the more severe the damage is, the more the tree should be influenced by the rut formation. However, the growth effects may also depend on the tree species due to differences in susceptibility (Wästerlund, 1983). A 10-cm deep rut formation may cut off

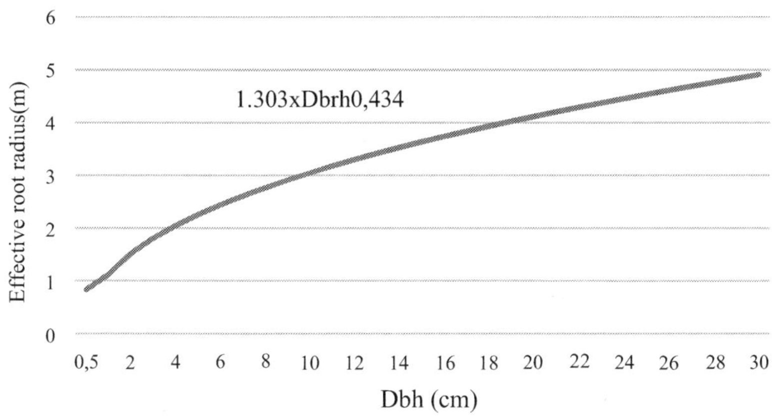

Figure 1.4 Model of the effective horizontal root spread, radius in meter. Single long laterals could, however, be longer, but on average root length could on medium soils follow the model. *(After Wästerlund, I., 1983. Growth reduction of trees near strip roads resulting from soil compaction and damaged roots. A literature survey.Sveriges Skogsvårdsförb. Tdskr., 81, 97—109. (Swedish, English summary).)*

or damage many of the feeding tree roots. Even a 6-cm rut may do harm. If the soil is not too moist, there will also be a compaction of the soil in the wheel rut.

Most likely, the regeneration of new roots will occur quite soon after a severe injury to the old root but the recolonization of the old territory may be impeded by soil compaction. A model for estimation of the growth reductions on average or poorer Scandinavian sites, based on the root growth model and literature data on growth reductions, is shown in Fig. 1.4. The model estimates the growth reductions as an average yearly reduction during 5 years after the damage. After that period, the growth is assumed to approach normal growth. Similar growth results have been obtained in Lithuania (Fig. 1.5).

According to experiments with uptake of labeled phosphate (see Chapter 4 and surveys of fruit bodies of mycorrhizal fungi (see chapter)), it could take 5 years or more until active tree roots are back in their former territory (Wästerlund, 1987,1989). Prediction models for the recovery time and such factors as climate and soil type do not appear to exist.

The growth effects probably depend on the interaction between soil compaction and root damage. Although there are no satisfactory experiments showing the effects of each type of damage, it appears that soil compaction may be more severe and longer-lasting type of damage than the root wounds (cf. e.g., Cochran and Brock, 1985).

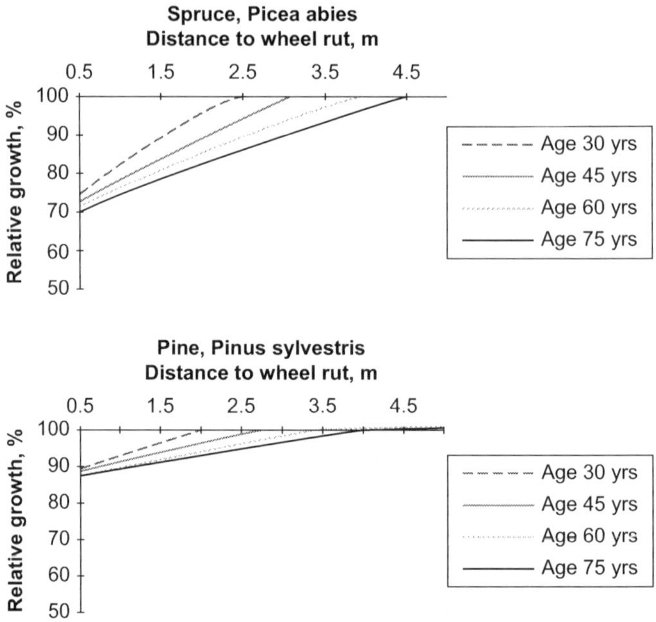

Figure 1.5 Growth reductions due to severe root and ground damage for Norway spruce and Scots pine according to a literature survey (Wästerlund, 1983).

Edge tree growth

A strip road in a newly thinned stand leaves an opening that in some cases is regarded as an unproductive area. The production loss is often solely blamed on that there are no trees in the opening and therefore no utilization of the soil. If this argumentation holds, then the unutilized resources ought to be used by herbs and bushes flourishing in this area. This is also seen in the area between deep wheel ruts but seldom in a natural opening in the stand.

On rich soils the trees are often competing for the light. On poorer soils they are also competing in the ground for nutrients. A 40-year-old tree may have an average horizontal radius of its root system of about 3 m (as seen above) but single roots may extend 5 or 10 m. The upper part of the ground is therefore almost completely interwoven by roots (Fig. 1.6).

The extra removal of trees to create space for the machines on the strip road will reduce the number of trees utilizing the ground. On the other hand, the remaining edge trees can explore that part of the ground and grow better.

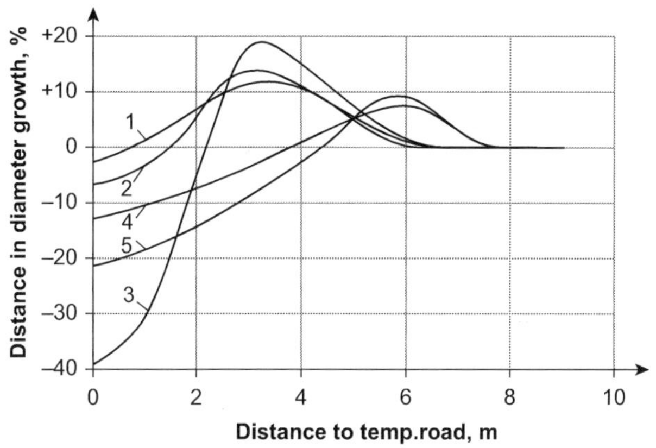

Figure 1.6 Diameter growth of the edge trees after skidding. 1: first thinning winter time with a 1.8 tonnes agricultural tractor (MT3). 2: first thinning winter time with a 4.5 tonnes crawler tractor (TAT-40). 3: first thinning summer time with TAT-40. 4: second thinning summer time with MT3. 5: second thinning summer time with TAT-40. *(After Šakunas, 1990, pers. com.)*

New information on the growth of undamaged edge trees is now available. A special topic in that context is comprised by line-thinning experiments. In one of them, two adjacent rows out of four were removed, leaving 4.5 m wide openings (Elfving, 1985). Four to seven years after thinning the line-thinned plots had only 4% lower growth compared to selective thinning. Isomäki (1986) studied the growth of edge trees 10 years after corridor cutting for electricity lines. He summarizes by saying that the extra growth of the edge trees beside, for instance, a 4-m wide corridor would equal the growth of a 2.3-m wide area (leaving 1.7 m "unproductive"). The edge trees were on average growing about 40% more in volume than trees further in. Similar basal area increment of edge trees was measured by Eriksson (1987). Thinned plots had lower volume production (7%—16%) than unthinned control, but the differences between 3.5 and 5 m wide strip road appear to be small. The results from studies of Niemistö (1989) and Pukkala (1989) indicate that a 2—3 m wide opening would give almost no growth reduction but 4—5 m (a. 17% treeless area) would give about 6% growth loss. The referred data concern 30- to 70-year-old stands, and in general it seems as if Norway spruce responds better to the opening than Scots pine do.

More data are needed to make good predictions on edge tree growth and the effect of openings on the total stand volume production. However,

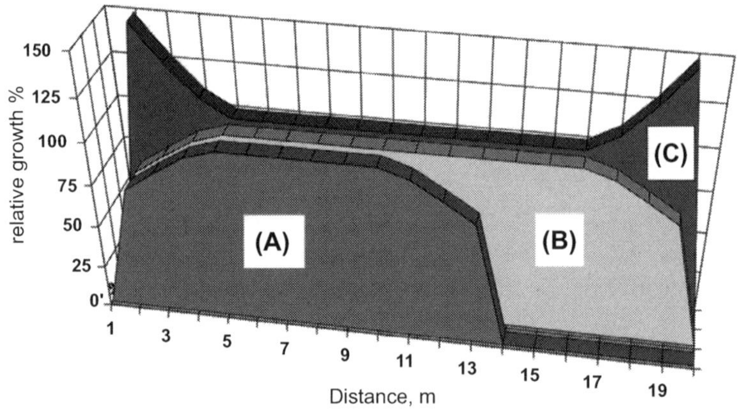

Figure 1.7 Schematic drawing of the growth of the stand after a thinning operation with 4 m wide roads. Alternative A (red (dark gray in print version)):5 m distance between roads and severe soil and root damage, 21% loss. Alternative B (green (gray in print version)); D:o but 20 m between roads, 15% loss. Alternative C (blue (black in print version)): D:o but no damage on the ground, 2% growth loss.

the referred data all show an extra growth of undamaged trees close to an opening that can be regarded as a yield on part of the opening (Fig. 1.7, alternative C).

This yield should also depend on the size of the tree root systems, e.g., the older the trees are, the deeper in the stand the growth effects might be expected according to the above-presented root growth model.

As a consequence of these new findings, my growth reduction model (Fig. 1.5) for damaged trees may be an underestimation of the growth reduction. Most of the data in that model are based on a comparison of the edge tree growth to those trees in the middle between strip roads. In reality, the growth of damaged edge trees (Fig. 1.7A and B) should be compared with undamaged edge trees (Fig. 1.7C). Then the real reduction of severely damaged edge trees could be in the magnitude of 50% and would then be similar to what is measured after clearfelling operations.

Suppose the width of the strip road is 4 m, the distance between these roads is 20 m and they have deep and compacted wheel ruts. The growth reduction could then amount to about 15% on areal basis and roughly the same in value (Fig. 1.7B).

However, the loss of the extra edge tree growth has to be added (road >2.5 m wide) amounting to about 5%–7%. Thus, the poor operation may roughly cost 20% of the production and be in the same magnitude as mentioned above for clearfellings. If now a light and small 2 m wide machine had been used instead and no damaged is caused on the ground, the machine is not even costing the production loss in the road! However, these arguments are not so good in late thinnings or selection thinnings because a bigger machine might be needed to handle the bigger trees. I think here lies the challenge for the future machine and technical development.

Spacing/cleaning operations

Mechanical cleaning of coppice in the young stands is introduced in some countries (e.g., Canada, France, and Sweden). When using the row cleaning method, more than half of the site could be passed over by the machine. Selective cleaning with a device mounted in the boom (used in Sweden) could reduce the influenced area to about 20% (Wästerlund, 1988). Studies concerning damage and growth reductions in such stands seem to be few and mainly be done in Sweden (Wästerlund, 1988,1990).

Damage surveys on the field experiments in planted Norway spruce showed rather high frequency of wounding, 6%—16%. However, it should be remembered that the machines were forced to work in stands that in many cases were not suitable for mechanical cleaning as a way to find out the limits of the method. Furthermore, it should be noted that all kinds of wounds were noted and maybe only half of the number could be considered as a serious damage. Four types of damage may be considered as severe (Wästerlund, 1990):

1. Driving over the seedlings with the wheel, which usually caused high mortality.
2. Straddling too high young growth with poor and/or sharp first contact between machine and tree may cause severe wounding.
3. Cutting the removed tree to high up. That will generally result in at least a double-topped tree with almost the same height as its neighbors. An instruction or an operators fault but nevertheless rather common.
4. Cutting in the tree with the cleaning device. A severe damage that was surprisingly uncommon.

A fifth type of damage occurred with the first machine type, a modified forwarder. Here straddled seedlings could be pushed out and squeezed between the bogie and the frame. The experiments started in the beginning of the 1980s with a modified forwarder—Bruunett Mini—and a disc with flails in the boom. The working method was to do selective cleaning but mainly between the rows, while the machine was continuously moving forward. Growth studies showed that seedlings standing close to the wheel ruts had a reduced growth extending for a period of 3—4 years. The growth reduction extended longer from the tracks for the 10- to 15-year-old stand compared with the 5-year-old stand (Figs. 1.8 and 1.9).

In 1987, a new type of cleaning machine was tested. It was a 5-tonnes 4-wheeled machine that later on came in a number of versions, one of them being a one-grip harvester. It was called "Lillebror" after the constructor. The working method was also changed to more effective selective cleaning/spacing, and therefore the machine was standing during work and in between just propelling forward to next working place. Growth measurements after operation with that machine with different wheel equipment showed that at least with 600 mm wide tyres no growth reductions on the edge trees were detectable. Smaller tyres gave some unsignificant indications on the growth.

Based on these studies, it was proposed that a 5- to 7-tonnes machine with a nominal ground pressure of 60—70 kPa and with low tractive forces could be an acceptable machine from a tree and soil damage point of view (Fig. 1.9).

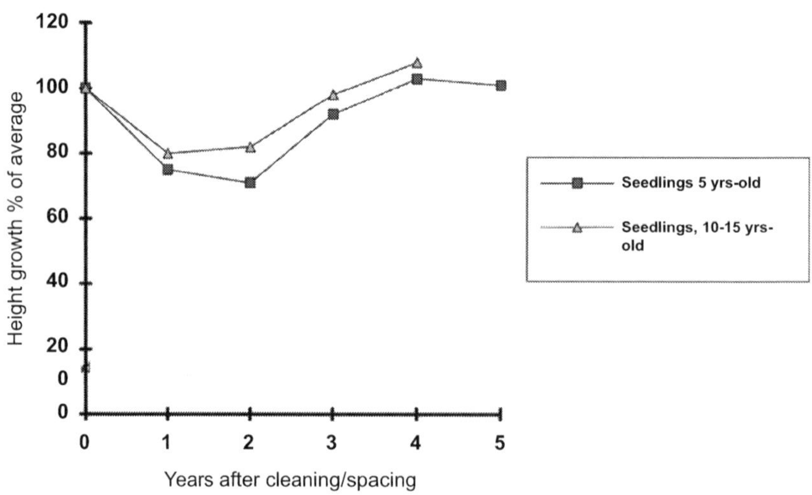

Figure 1.8 Height growth of otherwise undamaged seedlings close to the wheel ruts after mechanical cleaning in relation to growth of seedlings standing at least 2 m from the rut. Average growth for seedlings standing 0—0.5 m from the ruts (5-year-old seedlings) and 0—1 m (10- to 15-year-old seedlings (Wästerlund, 1990).

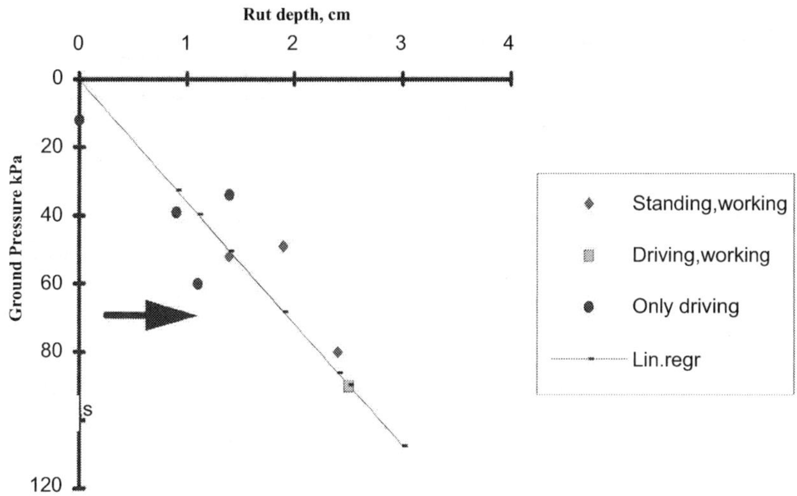

Figure 1.9 Rut depths at different maximum nominal ground pressure of the cleaning machines. The ground pressure is calculated as actual weight on the tire with the boom in working position divided by radius times the width of the tire. The *arrow points* on the ground pressure below which no negative effects were obtained. By coincidence almost the same as the foot print pressure for a man (Wästerlund, 1990).

Conclusions

Thus, in the good situation with tracks from clearfelling remoulded, the machine trails from thinning operations could usurp 15%–35% of the site area and half of that area could be markedly disturbed. In the heavy mechanized forestry with bad site preparation and mechanized spacing operations as well as thinning operations, the trails might cover nearly 100% of the area. Statistically 0.06% of the area could be run over 6 times and 9% zero times. In reality, the terrain will steer the paths to certain routes which will increase both of these figures.

The production losses due to soil and root damage justify the use of better methods and machinery and from an environmental point it is really needed. The mechanization of the forest work will continue and with more machine entries in the forest the forestry will be forced in that direction. However, there exist methods and machinery that at least in a first thinning operation may be acceptably friendly to the environment but the improvements must continue (Table 1.2).

The model in Table 1.2 was used for judgment of the degree of soil disturbance on the investigated thinnings and clearfellings. According to

Table 1.2 Proposed model for judgment of site damage after traffic Proposal of a model to judge the level of site disturbance due to traffic after clear-cutting.

	% share of the area with ruts			
Rut depth, cm	**<10%**	**10%–20%**	**20%–30%**	**>30%**
<5	No	No	Some	Great
5–9.9	No	Some	Great	Serious
10–14.9	Some	Great	Serious	Serious
15–19.9	Great	Serious	Serious	Unacceptable
>20	Unacceptable			

that model 45% of the harvested areas had "some" soil disturbance, whereas 55% had "no" soil disturbance. Still the question remains how to judge severe soil disturbance but on a very small part of the total area. That part has to be developed further and should also include some rules for the landings.

In the present movements in (and pressure on) the forestry toward sustainability and diversity, the harvesting methods will change toward more selective thinnings and selection cutting. Here, the handling of big trees in a growing stand with regeneration on the forest floor without damaging trees, plants, and soil is still a problem. I would say that the challenge for the 2030 is to improve methods and machinery for those types of operation.

References

Adams, P.W., Froehlich, H.A., 1981. Compaction of Forest Soils. Oregon State University, Extension Service, PNW-217.

Andersson, L., 1987. Influence of stem wounds on the growth of Scots pine. In: Knutell, H. (Ed.), Development of Thinning Systems to Reduce Stand Damages. Swedish University of Agricultural Sciences, Dept. of Operational Efficiency, Garpenberg. Res. Note No.98, pp. 53–63.

Anon, 1968. Glossary of terrain-vehicle terms. J. Terramechanics 5 (2), 53–69.

Benthaus, M., Matthies, D., 1993. Regeneration befahrenerWaldböden. Ein Vergleich von Breit- und Normalreifenspuren 6 JahrenachBefahrung. Allg. Forst. Z. 48, 448–451.

Biggs, A.R., Merrill, W., Davis, D.D., 1984. Discussion: response of bark tissues to injury and infection. Can. J. For. Res. 14, 351–356.

Björkhem, U., Lundeberg, G., Scholander, J., 1975. Root Distribution and Compressive Strength in Forest Soils. Royal College of Forestry, Department of Forest Ecology and Forest Soils, Stockholm. Res. Note.No. 22 (Swedish, English summary).

Bonneman, J., 1979. Untersuchungen über die Entstehung und Verhütung von "Wundfäulen" bei der Fichte. Forst. Fakultät Georg-August-Universität, Göttingen.Diss.

Burroughs Jr., E.R., Thomas, B.R., 1977. Declining Root Strength in Douglas-fir after Felling as a Factor in Slope Stability. U.S.D.A. Forest Service, Ogden. Res. Pap. INT-190.

Böhm, W., 1979. Methods of Studying Root Systems. Springer-Verlag, Berlin.

Cochran, P.H., Brock, T., 1985. Soil Compaction and Initial Height Growth of Planted Ponderosa Pine. U.S.D.A. For. Service, Portland, Oregon. Res. Note PNW-434.

Corns, J.G.W., 1988. Compaction by forestry equipment and effects on coniferous seedling growth on four soils in the Alberta foothills. Can. J. For. Res. 18, 75–84.

Coutts, M.P., 1977. The formation of dry zones in the sapwood of conifers. II. The role of living cells in the release of water. Eur. J. For. Path. 7, 6–12.

Campbell, R.G., Willis, J.R., May, J.T., 1973. Soil disturbance by logging with rubber-tired skidders. J. Soil Water Conserv. 28, 218–220.

Clayton, J.L., Kellogg, G., Forrester, N., 1987. Soil Disturbance - Tree Growth Relations in Central Idaho Clearcuts. U.S.D.A., Forest Service. Res. Note INT-372.

Clayton, J.L., 1990. Soil Disturbance Resulting from Skidding Logs on Granitic Soils in Central Idaho. U.S.D.A. Forest Service, Ogden. Res. Pap. INT-436.

Cullen, S.J., Montange, C., Ferguson, H., 1991. Timber harvest trafficking and soil compaction in western Montana. Soil Sci. Soc. Am. J. 55, 1416–1421.

Davis, S., 1992. Bulk density changes in two central Oregon soils following tractor logging and slash piling. West. J. Appl. For. 7, 86–88.

Dimitri, L., 1983. Die Wundfäule nach Baumverletzungen in der Forstwirtschaft: Entstehung, Bedeutung und die Möglichkeiten der Verhütung. Forstw. Cbl. 102, 68–79.

Elfving, B., 1985. Five year growth in a line-thinning experiment with pine and spruce. In: Gallagher, G. (Ed.), The Influence of Spacing and Selectivity in Thinning on Stand Development, Operations and Economy. Proc. IUFRO Project Group P4.02.02, Dublin, pp. 114–121, 1984.

Eriksson, H., 1987. New results from plot No. 5 at Sperlingsholm estate in southwestern Sweden in the European stem number experiment in *Picea abies*. Scand. J. For. Res. 2, 85–98.

Eriksson, L., 1981. Strip Roads and Damages Caused by Machines when Thinning Stands. Swedish University of Agricultural Sciences, Department of Operational Efficiency, Garpenberg. Report No. 137. (Swedish, English summary).

Eliasson, L., Wästerlund, I., 2007. Effects of slash reinforcement of strip roads on rutting and soil compaction. For. Ecol. Manag. 248, 118–123.

Freezailah, B.C.Y., 1994. Sustainability and tropical forests. ITTO Trop. For. Update 4 (2), 2.

Froehlich, H.A., 1978. Soil compaction from low ground-pressure, torsion-suspension logging vehicles on three forest soils. For. Res. Lab. Oregon State University, Corvallis. Res. Pap. 36.

Froehlich, H.A., McNabb, D.H., 1984. Minimizing soil compaction in the Pacific Northwest forests. In: Stone, E.L. (Ed.), Forest Soils and Treatment Impacts.Proc. Of Sixth North American Forest Soils Conference.University of Tennessee, Knoxville, June 1983. pp. 159-192.

Froehlich, H.A., Miles, D.W.R., Robbins, R.W., 1985. Soil bulk density recovery on compacted skid trails in central Idaho. Soil Sci. Soc. Am. J. 49, 1015–1017.

Froehlich, H.A., Miles, D.W.R., Robbins, R.W., 1986. Growth of yound Pinus ponderosa and Pinuscontorta on compacted soil in central Washington. For. Ecol. Manag. 15, 285–294.

Fröding, A., 1982. The Condition of Newly Thinned Stands. A Study of 101 Randomly Selected Thinnings. Swedish University of Agricultural Sciences, Dept. of Operational Efficiency, Garpenberg. Report No. 144. (Swedish, English summary).

Fröding, A., 1986. Damages - effects in the stand. In: Knutell, H. (Ed.), Development of Thinning Systems to Reduce Stand Damages. Swedish University of Agricultural Sciences, Dept. of Operational Efficiency, Garpenberg, pp. 42–55. Res. Note No. 98.

Fröding, A., 1992. Thinning Damage - A study of 403 Stands in Sweden in 1988. Swedish University of Agricultural Sciences, Department of Operational Efficiency, Garpenberg. Report No. 193. (Swedish, English summary).

Gent Jr., J.A., Morris, L.A., 1986. Soil compaction from harvesting and site preparation in the upper gulf coastal plain. Soil Sci. Soc. Am. J. 50, 443–446.

Gyldberg, B., 1993. Studies of the Workings of a Powered Disc Trencher. Swedish University of Agricultural Sciences, Department of Operational Efficiency, Garpenberg.Diss.

Hassan, A.E., 1978. Effect of mechanization on soils and forest regeneration. I. Coastal plain organic soils. Trans. ASAE 21 (6), 1107–1112.

Hatchell, G.E., 1981. Site preparation and fertilizer increase pine growth on soils compacted in logging. South. J. Appl. For. 5, 79–83.

Helms, J.A., Hipkin, C., 1986. Effects of soil compaction on tree volume in a California ponderosa pine plantation. West. J. Appl. For. 1, 121–124.

Hildebrand, E.E., Wiebel, M., 1986. Zur Bedeutung des Bodenwassergehaltes von Feinlehmstandorten bei der Entstehung von Bodenschäden durch Befahrung. Allg. Forst. Z. 41, 617–622.

Hofmann, R., 1989. Bodenschäden durch Forstmaschineneinsatz. Forstwissenschaftlichen Fakultät der Universität Freiburg i Br. Diss.

Incerti, M., Clinnick, P.F., Willatt, S.T., 1987. Changes in the physical properties of a forest soil following logging. Aust. For. Res. 17, 91–98.

Isomäki, A., 1986. Effects of line corridors on the development of edge trees. Folia For. 678.

Isomäki, A., Kallio, T., 1974. Consequences of injury caused by timber harvesting machines on growth and decay of spruce (*Picea abies* (L.)Karst.). Acta For. Fenn. 136.

Ivanov, B.N., 1976. Dynamics of density of soil in clearings of spruce forests. Lesovedenie (4), 2630 (Russian, English summary).

Jakobsen, B.F., 1983. Persistence of compaction effects in a forest kraznozem. Aust. For. Res. 13, 305308.

Kardell, L., 1978. Increment losses of Norway spruce caused by tractor logging. Sveriges Skogsvårdsförb. Tdskr. 76, 305–322 (Swedish, English summary).

Karr, B.L., Hodges, J.D., Nedbecker, T.E., 1987. The effect of thinning methods on soil physical properties in north-central Mississippi. South. J. Appl. For. 11, 110–112.

Kreh, R.E., Burger, J.A., Torbert, J.L., 1985. Soil Compaction From Tracked and Rubber-Tired Tractors and Its Influence on Seedling Survival and Growth. U.S.D.A. General Technical Report Forest Service Southern Forest Exp. Stn. SO-54, pp. 327–330.

Lee, H.W., Johnson, L.R., Stoker, J., 1990. Soil compaction from mechanical fellers. In: Focus on Renewable Natural Resources. University of Idaho, Forest, Wildlife and Range Exp. Stn., Moscow, Idaho: The Station. V.15, pp. 25–26.

Lenhard, J.R., 1986. Changes in void distribution and volume during compaction of a forest soil. Soil Sci. Soc. Am. J. 50, 462–464.

Lull, H.W., 1959. Soil Compaction on Forest and Range Lands. U.S.D.A., Forest Service, Washington (Misc. publ).

Malmer, A., Grip, H., 1990. Soil disturbance and loss of infiltrability caused by mechanized and manual extraction of tropical rainforest in Sabah, Malaysia. For. Ecol. Manag. 38, 1–12.

Martin, C.W., 1988. Soil disturbance by logging in New England - review and management recommendations. North. J. Appl. For. 5, 30–34.

McCreary, D.D., Perry, D.A., 1983. Strip thinning and selective thinning in Douglas-fir. J. For. 81, 375377.

McLaughlin, J.A., Pulkki, R.E., 1992. Assessment of wounding at two commersially thinned Jack pine sites. North. J. Appl. For. 9, 43–46.
McNeel, J.F., Ballard, T.M., 1992. Analysis of site stand impacts from thinning with a harvester-forwarder system. J. For. Eng. 4, 23–29.
Meng, W., 1978. Baumverletzungendurchtransportvorgängebei der Holzernte. Schrftr. d. Landesforstverw, Baden-Würtemburg, 53.Diss.
Murphy, G., 1982. Soil damage associated with production thinning. N. Z. J. For. Sci. 12, 281–292.
Myhrman, D., 1990. Factors influencing rut formation from forestry machines. In: Proc. 10th International Conference of the ISTVS, Kobe, Japan: 467-475.
Niemistö, P., 1989. A simulation method for estimating growth losses caused by strip roads. Scand. J. For. Res. 4, 203–214.
Nilsson, P.-O., Hyppel, A., 1968. Studies on decay in scars of Norway spruce. Sveriges Skogsvårdsförb. Tdskr. 66, 675–713 (Swedish, English summary).
OLoughlin, C.L., 1974. A study of tree root strength deterioration following clearfelling. Can. J. For. Res. 4, 107–113.
OLoughlin, C.L., Watson, A., 1979. Root-wood strength deterioration in radiata pine after clearfelling. N.Z. J. For. Sci. 9, 284–293.
OLoughlin, C.L., Watson, A., 1981. Note on root-wood strength deterioration in Nothofagus Fusca and N. truncata after clearfelling. N. Z. J. For. Sci. 11, 183–185.
Olsen, H.J., Wästerlund, I., 1989. Terrain and Vehicle Researchs with Reference to Forestry at Swedish University of Agricultural Sciences. SLU; Department of Operational Efficiency, Garpenberg. Uppsatser&Resultat No 149. (in Swedish with English summary).
Pukkala, T., 1989. Methods to describe the competition process in a tree stand. Scand. J. For. Res. 4, 187202.
Reisinger, T.W., Simmons, G.L., Pope, P.E., 1988. The impact of timber harvesting on soil properties and seedling growth in the South. South. J. Appl. For. 12, 58–67.
Reisinger, T.W., Pope, P.E., Hammond, S.C., 1992. Natural recovery of compacted soils in an upland hardwood forest in Indiana. North. J. Appl. For. 9, 138–141.
Schäfer, T., Sohns, D., 1993. Minderung der Bodenverdichtung durch eine Reisigauflage. Allg. Forst. 48, 452–455.
Schäffer, J., Hildebrand, E.E., Mahler, G., 1991. Bodenverformung beim Befahren.Wirkung der Armierung durch Reisigmatten. Allg. Forst. Z. 46 (11), 550–554.
Shetron, S.G., Sturos, J.A., Padley, E., Trettin, C., 1988. Forest soil compaction: effect of multiple passes and loadings on wheel track surface soil bulk density. North. J. Appl. For. 5, 120–123.
Shigo, A.L., 1985. Compartmentalization of decay in trees. Sci. Am. (April-85), 76–83.
Shishiuchi, M., Adachi, K., 1982. Influence of tractor logging on soil surface condition. (I) Effect of soil compaction from tractors on the growth of planted Japanese larch seedlings. J. Jap. For. Soc. 64, 136–142.
Smith, R.B., Wass, E.F., 1979. Tree Growth on and Adjacent to Contour Skidroads in the Subalpine Zone, Southeastern British Columbia. Environment Canada, Canadian Forestry Service, Pacific Forest Research Centre, Victoria. BC-R-2.
Snider, M.D., Miller, R.F., 1985. Effects of tractor logging on soils and vegetation in eastern Oregon. Soil Sci. Soc. Am. J. 49, 1280–1282.
Stewart, R., Froehlich, H.A., Olsen, E., 1988. Soil compaction. An economic model. West. J. Appl. For. 3, 20–22.
Turmanina, V., 1965. [The strength of tree roots]. Bjull. Mosk. Obš. Ispyt. Prir. (Otd. Biol.) 70 (5), 36–45 (Russian).
WCED (World Commision on Environment and Development), 1987. Our Common Future. Oxford University Press, Oxford.

Wert, S., Thomas, B.R., 1981. Effects of skid roads on diameter, height and volume growth of Douglasfir. Soil Sci. Soc. Am. J. 45, 629–632.

Williamson, J.R., 1990. The Effects of Mechanized Forest Operations on Soil Properties and Site Productivity. Tasmanian Forest Research Council, Inc., Hobart. Research Report No. 5.

Wilpert, K., Parbs, J., 1992. Flächenbedeutung von Bodenschädenbei der Erstforstung mit kombiniertem Miniharvester - forwarder - Einsatz. KWF. Forsttechnische Informationen 44 (4), 25–28.

Wronski, E.B., 1984. Impact of tractor thinning operations on soils and tree roots in a karri forest, western Australia. Aust. For. Res. 14, 319–332.

Wästerlund, I., 1983. Growth reduction of trees near strip roads resulting from soil compaction and damaged roots. A literature survey.Sveriges Skogsvårdsförb. Tdskr. 81, 97–109 (Swedish, English summary).

Wästerlund, I., 1986. The Strength of Bark on Scots Pine and Norway Spruce Trees. Swedish University of Agricultural Sciences, Department of Operational Efficiency, Garpenberg. Report 167.

Wästerlund, I., 1987. Three case studies about growth effects after soil compaction and root damages. In: Knutell, H. (Ed.), Development of Thinning Systems to Reduce Stand Damages. Swedish University of Agricultural Sciences, Department of Operational Efficiency, Garpenberg, pp. 41–52. Res. Note No. 98.

Wästerlund, I., 1988. Damages and growth effects after selective mechanical cleaning. Scand. J. For. Res. 3, 259–272.

Wästerlund, I., 1989. Effects of damage on the newly thinned stand due to mechanized forest operations. In: Seminar on the Impact of Mechanization of Forest Operations on the Soil. Proc. ECE/FAO/ILO Seminar, Louvain-la-Neuve, Belgium, pp. 165–175.

Wästerlund, I., 1990. Damage to the Ground and the Stand after Mechanized Cleaning. Swedish University of Agricultural Sciences, Department of Operational Efficiency, Garpenberg. Res. Note No 193. (Swedish, English summary).

CHAPTER 2

Impacts of soil disturbance on forest and forest soils

Introduction

In the previous chapter, I was discussing some results and theories about the effect of soil disturbance on forest growth in a long perspective as we can measure or see it on the stem growth. However, there are several aspects on the soil environment that interact to affect the ecosystem. In field conditions, these factors are often very dynamic and can seldom be determined, with a high degree of certainty, which factor was dominant. Trees are also relatively long-living crops, which may react dynamic on the impulse and making single observations difficult to evaluate. These aspects will be discussed below in form of root dynamics, microbial reactions, as well as nutrient situation and erosion risks which may influence the water quality. The review is mainly restricted to results obtained in field experiments and observations.

Root dynamics

The study of roots and root systems is fraught with difficulties (Sutton, 1991). Foremost among these are the variability of soil on every scale from micro to global and the variability of root system (Sutton, 1991). Taylor and Klepper (1978) listed 24 soil factors that affect rooting to illustrate the almost overwhelming complexity of soil—root system. To exemplify that I will start with the photosynthesis and give examples of how the input energy is used within the tree and what goes below the ground (cf. Fig. 2.1).

Only a minor part of the photosynthetic capacity in a tree is used for stem wood production. Measurements on a 14-year-old Scots pine (*Pinus sylvestris* L.) indicated that the utilization of carbon in the tree was distributed between respiration (10%), stem growth (9%), branch growth (8%), current needle

Soil and Root Damage in Forestry
ISBN 978-0-12-822070-2
https://doi.org/10.1016/B978-0-12-822070-2.00002-X
© 2020 Elsevier Inc.
All rights reserved.

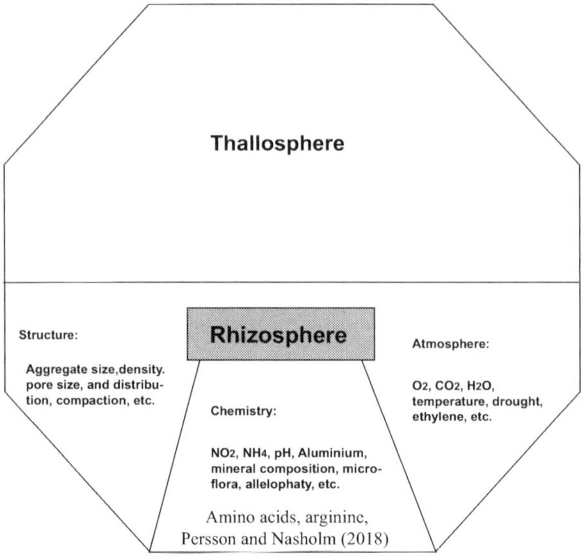

Figure 2.1 Diagram representing the main environmental factors which interact with the plant rhizosphere to control (modulate) plant growth and development Amino acids, arginine, Persson and Nasholm (2018). *(Modified from Zobel, R.W., 1992. Soil environment constraints to root growth. In: Hatfield, J.L., Stewart, B.A. (Eds.), Limitations to Plant Root Growth. Advances in Soil Science, vol. 19. Springer-Verlag, New York. pp.27—52.)*

growth (17%), and fine root growth (57%) according to Ågren et al. (1980). Similar results are also obtained for 18- to 115-year-old Norway spruce trees (*Picea abies* (L.) Karst.), indicating that 41%—74% of the net photosynthesis disappears as respiration and a sink to the roots (Tamm, 1974; Schulze et al., 1977). Trees are not unique in that respect as 47% sink to the roots in legumes have been reported (Minchin and Pate, 1973). In an experiment with Scots pine with irrigation and fertilization, only 40% of the net photosynthesis was used below ground, while it was 65% for the control trees (Ågren and Axelsson, 1980). In the latter study, the respiration losses were 10% in both cases and thus the fertilized and irrigated trees used more of the net photosynthesis to produce needles and stem wood. This shows (i) a major part of the net photosynthesis is used below ground and (ii) at richer conditions trees have a priority for above-ground production, while at poor soils the search for nutrients and water have to be preferred.

Each year there is a considerable growth of new roots (Persson, 1980). The poorer the soil, the more roots are formed, and the more superficial the new roots become (Meyer, 1967). However, the root extension is not running very fast in elderly trees, the new roots are filling out the soil between other roots, and a great part of the fine roots are just replacing the old dying roots. The necromass of roots <2 mm in diameter may be 30%–50% of the new production (Persson, 1980). Thus, there is a dynamic process going on below ground that can help the tree to adapt to changing soil conditions.

Above I called the part of the net photosynthesis transferred below ground as a "sink," because there are other sources of reduction of C in the ground than the root production. Root exudates are the other part of the sink. Carbohydrates, amino acids, organic acids, enzymes, and other compounds are exudated to stimulate or inhibit fungi, bacteria, or nematodes (Russell, 1977). They have shown to exist for monocotyledons, dicotyledons, and gymnosperms. The amounts of material lost through the roots under natural conditions are unknown, because it is virtually impossible to measure it undisturbed. Estimations indicate that the exudation accounts for up to 10% of the total carbon assimilated by the plants (Russell, 1977). Since tree roots almost always are associated with mycorrhizae in the nature, it may be reasonable to assume that this is a major feeding source for the mycorrhizae. Calculations by Romell (1939) based on the fruit body production of mycorrhizal mushrooms indicate that the fungi may use up about 400 kg food per ha, which may be in the same order as the estimated of amount of root exudation. Other aspects of mycorrhizae will be dealt with below.

Thus, the development of the tree roots is a long-term dynamic process with a number of factors influencing the root growth and soil fertility is one of more important factors.

Microbial interactions

The forest floor layer (definition in Jenny, 1980) and the associated microflora and fauna are probably the most dynamic phase of the forest ecosystem and are of great importance to several aspects of forest land management (Woolridge, 1970) and their widespread mycelium (hyphae). The mycorrhizae are a symbiosis between the trees and associated mushrooms. Both partners are gaining on the association. Trees are feeding the fungus as mentioned above and in return the fungus is helping the tree in the uptake of nutrients and water. Often a distinction is made between ectomycorrhizae and endomycorrhizae. Ectomycorrhizae is common among conifers and the fungus does not enter inside the root cortex cells. Vesicular-arbuscular mycorrhiza (VAM) is one type of endomycorrhizae entering inside the cortex cells in the roots. Simmons and Pope (1987) made some studies of the effect of VAM on the root growth of yellow poplar (*Liriodendron tulipifera* L.) and sweet gum (*Liquidambar styraciflua* L.) with two fungi species. Results indicate that for yellow poplar, which has a coarse root system, soil compaction effects may outweigh mycorrhizal benefits at higher bulk densities. On the other hand, the results for sweet gum with a fibrous (fine and highly branched) root system suggest that effects of soil compaction can be alleviated by inoculum with one of the mycorrhizal fungi.

The visibility of the fine tree root damage is rather difficult. Wästerlund (1989) used the appearance of fruit bodies of mycorrhizal mushrooms after a thinning operation in a spruce forest, to show the rate of root reoccupation of the area between wheel tracks. He observed that it took 5 years until the fruit bodies showed up in the middle of the strip road (Fig. 2.3). One aspect of that experiment may be that the mycorrhizal fungi could be helping the tree roots to find their way back to former territory.

The other aspect when looking at microbes associated with trees is to evaluate the effect of soil compaction on pathogens of tree roots. In one experiment, Smeltzer et al. (1986) found a reduced number of bacteria and soil fungi in the compacted soil compared to the control, except for *Fusarium* spp—a fungal genus with many plant pathogenic species—which increased considerably. The difference existed 1 and 2 years after treatment, but not 5 years later. Otherwise, there is very little information on this aspect although there might be some foreseen literature dealing with it concerning cultivation of fruit trees and cotton. We know that in general the rhizosphere flora is protecting the tree roots (cf. Russell, 1977) and part of the exudates may be considered as a pay-off for the "mafia." When a wheel or a track is crushing or otherwise damaging the superficial roots beneath it in the soil, the flow of exudates out to these organisms is strangled. A new competitive situation is created, and I believe in this situation pathogens may take their chance. Besides of that, in the damaged root may start to leak products from the wound and those wound products may attract the pathogens. In a south Swedish study, it was found that the frequency of root rot attacks (*Heterobasidion annosum* (Fr.) Bref.) was very high near the strip road after three to four thinning operations, and decreased very fast with increased distance to the strip road (Fig. 2.2).

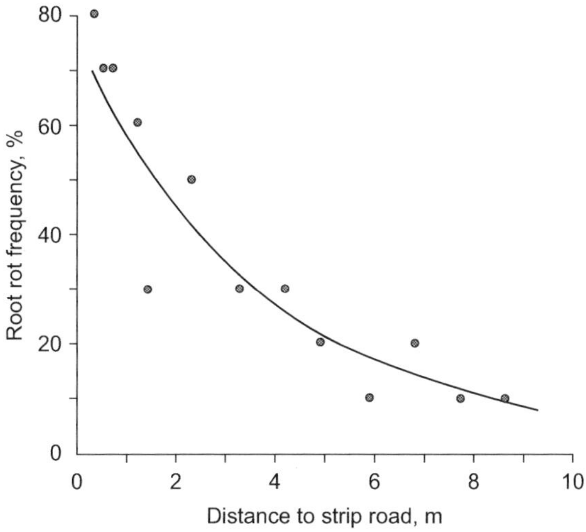

Figure 2.2 Average distance to strip road and root rot frequency. Average for three newly thinned Norway spruce stands thinned for the third or the fourth time at the age 60 year, Biskopstorp. 10 trees per group. *(Redrawn from Bernhardsson, A., Martinsson, L., 1986. Rötfrekvens i högproduktiva granskogar. [Root rot frequency in highly productive spruce forests]. Swedish University of Agricultural Sciences, Department Forest engineering school, Skinnskatteberg. Diploma work. (Swedish).)*

Root damage from the 12–35 cm deep wheel ruts may have been a trigger for the fungus, but the study does not show why the root rot was so frequent near the strip road (Bernhardsson and Martinsson, 1986).

Soil fauna

The number of individs in the forest soil fauna can be considerable and variable. Lohm (1977) showed that even in a pine stand on a dry Swedish sandy site, a foot print (size 44, english size 10, multiply with 40 to get numbers per m^2) may contain 400 *Enchytraeidae*, 1500 *Collembola*, 20 *Protura*, 15 *Coleoptera* (beetles), 70 *Diptera* (flies), 10 *Aranae* (spiders), 19,000 *Acari*, and 100,000 *Nematoda*. In a nitrogen fertilization experiment, it was shown that the occurrence of *Enchytraeidae*, *Collembola*, and *Cryptostigmata* showed (i) a short-term effect with decreased abundance owing to ammonium toxicity or a "salt effect" and (ii) a long-term effect with increasing abundance resulting from increased food supply (Lohm et al., 1977). In one experiment, Smeltzer et al. (1986) found a severely reduced number of soil macroarthropods and nematodes 1 year after a compaction

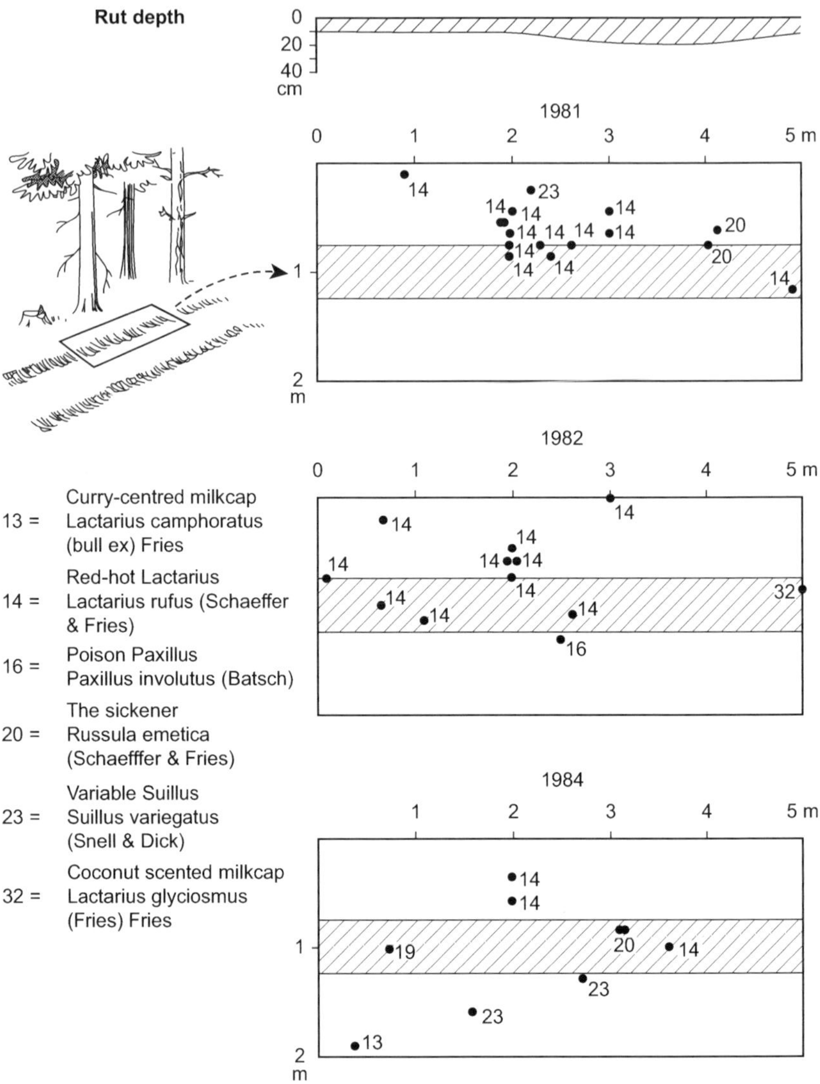

Figure 2.3 The occurrence of fruit bodies in the middle of the strip road the years 1981–84 after a thinning operation in 1979 creating 20–25 cm deep wheel ruts (Wästerlund, 1989). *(English names after Schalkwijk-Barendsen, H.M.E., 1991. Mushrooms of Northwest North America. Lone Pine Publishing Auburn WA and Watling, R., 1973. Identification of the Larger Fungi. Hulton Group Keys. Amersham.)*

treatment, but 4 years later the populations were about normal. Otherwise, effects on soil fauna due to traffic are poorly studied. The examples show, however, that changes in the soil fauna populations may be rapid.

Rooting behavior

Hartmann (1951) noted the relationship between site fertility and root branchiness; long and seldom-branched lateral roots indicate soils of low fertility, whereas compact complexes of fine roots indicate high fertility. Voronkov (1974) found that the root systems of highly productive Scots pine forest stands penetrate deeper into the soil and occupy a larger area than the root systems of low site class stands. Binkley (1986) hypothesized that a low level of nutrient availability might require the development of a larger root system than under conditions of higher nutrient availability. However, still a great portion of the fine root system may be in the uppermost 10 cm of the profile almost regardless of the soil type, as was found for Scots pine and Norway spruce in Germany (Meyer, 1967; Scherfose, 1990; Rastin, 1991). It is more pronounced on poor and wet soils. Only in very rich Braun soils the fine roots had a more even distribution with depth. The turn-over in organic layers attracts roots and scalping the upper soil layers causes a great increase of root growth (Bilan, 1960). A regeneration of a damaged seedling (*Abies procera* Rehd.) lateral root may occur within 20–30 days during the growing season (Wilcox, 1955). In a dormant root, the necrotic zone has a much greater extent than in active roots. The effects of wounding were found below the pruned surfaces at a distance equal to approximately twice the root diameter. Extensive root pruning may retard the development of new roots (Wilcox, 1955).

Thus, a deep rutting in a growing stand on a poor site may cause more root damage and have a greater impact on the tree growth than on rich soils and the response may also depend on the season the damage was made.

When talking about rooting depths, there seems to be three major factors controlling it. One is the genetic inheritance (tree species) and the second one is the climate type. The third one, soil properties, is the location of available nutrients and the possibilities for the roots to reach them, e.g., the occurrence of a hardpan may impair the root penetration (cf. Forristall and Gessel, 1955; Scherfose, 1990). Many broad leaved trees seem in general to have a very deep root system, e.g., beech (*Fagus sylvatica* L.) and oak (*Quercus* spp.), whereas spruce trees (*Picea* spp.) are often having a shallow root system. Spruce seems in generally well able to adapt their root systems to new conditions imposed on the roots (Sutton, 1991). But even species that develop strong tap roots, the form a that root system is often modified very early by injury or an adverse soil factor. With increasing maturity or size, the tree root systems become increasingly dominated by soil conditions, the more so as these depart from the ideal soil, freely penetrable by roots (Sutton, 1991). Schuster (1936) made this statement after studying root development in orchard soils in Oregon: "*The depth of root penetration ... is believed to be more often controlled by the characteristics of the soil than by the characteristics of the tree itself. In other words, the ideal soil is too often lacking ... and only in ideal soil do the individual characteristics of the plant express themselves fully.*"

If the soil is stony and the climate is humid, most of the nutrient turnover is taking place superficially, and hence, most of the roots are found in the top soil (cf. review by Wästerlund, 1983). In prairie soil roots were found down 4 m depth (Yeager, 1935) but in deep sandy soils pine roots are found down to 3 m depth (Laitakari, 1927; Vater, 1927). In such areas, it is essential for the water supply of the tree that the roots go deep. A compacted layer in the top soil may prohibit deep root penetration and thus cause a severe water deficit for those trees. A hypothesis could be that in areas with summer drought the need for deep rooting is more essential and soil compaction may cause a greater impact than at more humid conditions.

Conclusion: Both the nutrient status and nutrient distribution in the soil and the availability of water determine if trees need to be deep rooted to have a continued good growth after a soil disturbance.

Root strength

There exists quite little information about the root strength, especially for European conditions. In a study in Oregon, USA, (Burroughs and Thomas, 1977), it was shown that trees in the mountain areas had a much lower root strength than at the coastal area. The expectation would be that coniferous trees produce less strong root wood on richer sites compared with poor sites. A such result was obtained in Sweden (Wästerlund, see Chapter 5. Fig. 2.4) where Norway spruce trees on a rather fertile agricultural site had 35% lower tensile strength than at a poorer forest site. In the same way the elasticity of the roots was also reduced with 35%. The regulation of growth factors controlling root growth is much less understood than those for shoots. It appears that they are affected by the

Figure 2.4 Roots below a wheel rut. The coarser long lateral has actually bark peel off on the upper side. 10 cm depth in the humus layer. *(Photo I. Wästerlund.)*

environment of the roots tips (Sutton, 1991). Large-diameter vessels in angiosperm roots tend to occur in fertile, well-cultivated, periodically moistened soil. In contrast, small-diameter vessels tend to occur deep in the soil, where growth is physically restricted, where nutrition is poor, and where water supply is constant or excessive and probably associated with poor aeration (Fayle, 1980) (see also Chapter 5).

Thus, although it might not be seen as an effect on the tree growth, a deep wheel rut might cause more physical root damage on the fertile site than on the poor site due to the softer roots on the fertile site (e.g., at a thinning operation).

Soil strength

Compaction of the soil means increased soil strength and increased penetration resistance for the roots. Sands et al. (1979) found a cone resistance of above 3000 kPa to be the upper limit for a pine root to penetrate in sandy soils. However, both the soil type and the moisture content in the soil influence the ability for roots to penetrate hard soils. Russell (1977) has made a compilation of different soil strength data based on cone resistance values and found a striking convergence with soil strength measured as cone resistance (diam. 10 mm) and root penetration regardless of soil type (Fig. 2.6). Similar results have been shown for different types of agricultural crops (Bennie, 1991). In general, soil strength above 2000 kPa seems to severely impair root growth, but some species may become impeded already at much lower values. However, the levels may also depend on the size of the penetrometer cone used. As a comparison, it may be noted that measured root growth pressures of several crops range from 700 to 2500 kPa according to a review by Sharpley et al. (1992).

Compared with data concerning maximum cone penetrometer resistance, the maximum bulk density values for root penetration show a much greater variability (from 1.4 to 1.85 $Mg\,m^{-3}$, Fayle and Pierpoint, 1975; Russell, 1977; Sharpley et al., 1992) indicating the bulk density values to be a rather bad measure in this respect and very dependent on soil types. Douglas-fir (*Pseudotsuga menziesii* (Mirb.) Franco) seedling root penetration was prevented at bulk densities varying from 1.74 to 1.83 $Mg\,m^{-3}$ depending on soil type (Heilman, 1981). Douglas-fir, red alder (*Alnus rubra* Bong.), and lodgepole pine (*Pinus contorta* Douglas) seem to be less susceptible to soil compaction than, e.g., Sitka spruce (*Picea sitchensis* (Bong.) Carr.) and western red cedar (*Thuja plicata* Donn) (Minore et al., 1969). On the other hand, bulk density may be an important parameter for the determination of the bearing capacity of a soil (cf. Jakobsen and Greacen, 1985). See further next chapter.

Russell (1977) is taking up the ongoing debate about if root penetration may be impeded by an increased soil strength or if it is only a result of poor aeration. In reality, it may be a combination of both these factors (Fig. 2.5). Elastic changes in the soil might result in changes in the pore size distribution in spite of unaltered bulk density (Lenhard, 1986). In a comparative study, Gilman et al. (1987) found compaction causing significant shallower roots in both normal and reduced oxygen areas (14.4%–14.9% O_2). In areas with reduced oxygen but normal bulk density, the rooting was not shallower than in normal oxygen areas.

Agricultural crops (see also Fig. 2.6) seem to have ceased root growth when the oxygen content goes down below 10% (Russell, 1977). Root elongation of herbaceous plants slow down when the oxygen concentrations fall below about half the atmospheric concentrations (Jackson and Drew, 1984). In general, it is accepted that an oxygen level around 12%–15% in the soil is restricting the root growth (cf. Gilman et al., 1987). The other factor is the oxygen diffusion rate through the soil, which among other things is depending on the metabolic activities in the soil, e.g., microbial activities. However, we should remember the high variability in oxygen demands among tree species and that the need for oxygen is also depending on the present growth rate. According to Greenwood (1969), the metabolic activities of mustard (*Brassica* sp. L.) roots were unaffected by lower oxygen concentrations at the root surface until the concentration

(A)

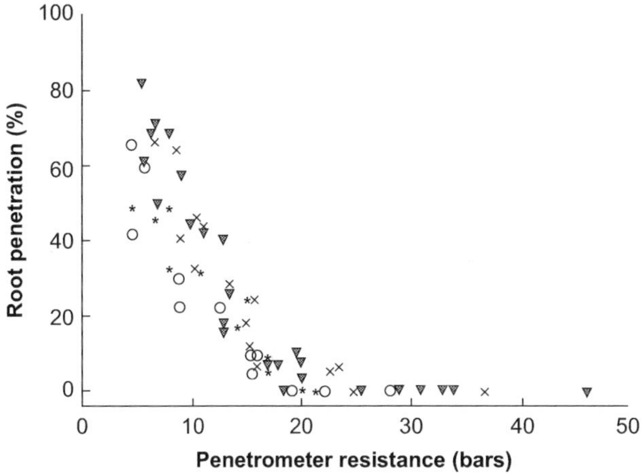

(B)

Mechanical analysis of soils (%)

	Sand	*Silt*	*Clay*
	> 50 μ	*2–50 μ*	*< 2 μ*
×	44	37	19
▼	73	20	7
*	80	11	9
○	83	8	9

Figure 2.5 Relationships between root penetration and penetrometer resistance in four contrasting soils. *(From Russell, R.S., 1977. Plant Root Systems. Their Function and interaction with the Soil. McGraw-Hill Book Company Limited, London.)*

became extremely low. He criticized a number of other experiments that seemed to show relatively high oxygen partial pressures impeding root elongation because no attempt had been made to minimize the path lengths of oxygen diffusion through the aqueous phase in those experiments. Greenwood (1969) argued that "*as far the effects of aeration on root growth are concerned, the sole interest is in the extent of oxygen-free zones in the soil.*" The extent of such zones is influenced enormously by the distribution of water in the soil, which in turn is highly dependent on the soil pore system (Sutton, 1991). With a given amount of gas-filled pore space, the smaller the pores, the more effective the aeration; with larger pores, the area of interface is smaller and the extent of anaerobic zones larger (Sutton, 1991). The question is how this holds on a compacted soil, since a compaction tends to make the pore distribution extending mainly in a horizontal way

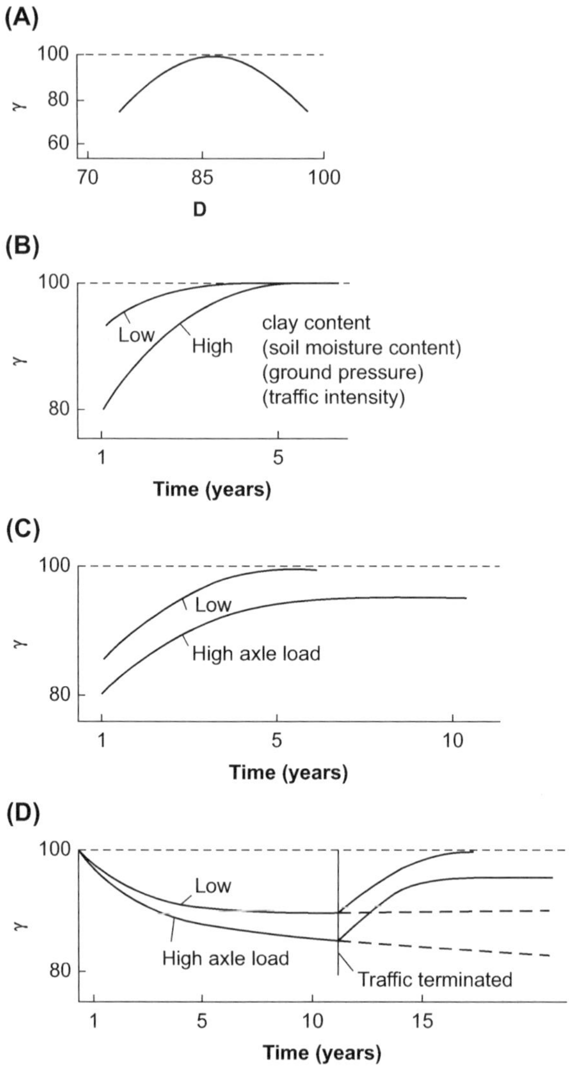

Figure 2.6 (A) Yield of spring barley under normal Swedish weather conditions as a function of the degree of compactness (Y and D) of the plow layer. Degree of compactness is comparison between undisturbed and compacted from 70 to 110 cm level. (B) Residual effects of compaction of the plow layer on one occasion, shortly before ploughing, as a function of clay content, soil moisture content at time of traffic, and ground pressure or traffic intensity. Annual ploughing is assumed. (C) Residual effects of intensive traffic on one occasion as a function of the axle load of the vehicles. Annual ploughing is assumed. (D) Cumulative effects of annually repeated traffic of moderate intensity as a function of the axle load of the vehicles used. The residual effects after terminating the traffic are also indicated. Annual ploughing is assumed. *(After Håkansson, I., 1990b. Soil compaction control - objectives, possibilities and prospects. Soil Technol. 3, 231–239.)*

(Olsson, 1986) and thus impeding the aeration of the soil. Further discussion about the measurement of the soil aeration status can be found in Stolzy et al. (1981).

In times with low growth activity, trees can withstand flooding for quite a long time if the temperature is low. For a wide range of tree species, both conifers and deciduous, the duration of flooding is critical only during the growing season (Gill, 1970). Coutts and Philipson (1978) found seedling roots of lodgepole pine (*P. contorta* Douglas) to grow more than 100 days in waterlogged soil at a temperature of 10°C but at 20°C the root growth had ceased at that time and the penetration depth was only half (100 mm below water table) as much as the one at lower temperature. Roots of Sitka spruce (*P. sitchensis* (Bong.) Carr.) grew hardly into the waterlogged soil in the same experiment. It was found that roots of lodgepole pine growing in the water had large and long extending air-filled spaces in the stele (aerenchyma?). However, in both species, mycorrhizae were abundant but were present only above the water table. Maybe spruce trees are more impeded by soil compaction than pines due to higher oxygen demand in the roots?

According to Vomocil (1965), an air-filled porosity near or below $0.10\ m^3\ m^{-3}$ (= macropores) is considered to be limiting the plant growth as a result of poor aeration. Heilman (1981) found a pore space of 30%–27% to be restricting for Douglas-fir seedling root growth. On the other hand, we should also remember that the temperature differences between soil and air is a driving force for the gas exchange and in many reports this aspect is not mentioned (measured?) (cf. Jaynes, 1990).

As pointed out before, we must consider the high variability and the dynamics of root growth. Nothing is really static and as soon as cracks, worm channels, etc., are formed and the soil opens, there is another situation (cf. Ehlers, 1975). Nambiar and Sands (1992) showed in a very good experiment how compaction changed the rooting behavior (impeded downward growth) of radiata pine (*Pinus radiata* D. Don.), and how that affected the water balance of the trees during the summer, which influenced tree growth (height, diameter, and taper). However, a perforation down to 1 m rendering 0.2% of the soil volume into vertical channels practically nivellated the impact of soil compaction. Such channels may also be formed in the soil through the decay of old roots (Gaiser, 1952). Roots had a remarkable ability to find these vertical channels and utilize them for the search of water in deeper soil layers.

Different species act different on compacted soil. Spruce trees seem to accept the situation and just grow on the top soil. Others strive for getting down to really reach the water supply and may find different types of channels to pass through. One way of improving the situation may be to sow out species growing on moist soils like *Juncus* L. Klimo (1983) showed that *Juncus conglomeratus* L. is extremely tolerant to compacted soil and is often the first colonisator. I have observed that *Juncus effusus* L. also is common on those spots. Klimo (1983) noted that when the soil was colonized by *J. conglomeratus*, the soil was successively loosening by its roots.

In agriculture, the Håkansson concept about degree of compactness(Y) is being accepted. He introduced soil testing with help of an oedometer in which big soil samples from the field (300 mm × 100 mm) are loaded with a 200 kPa static load till consolidation (2–4 days). Then soil volume is measured and the soil cores are weighed both wet and dried. The procedure gives a very reproducible measure of a maximum compactness level (Håkansson, 1990a). In a number of field experiments, it has been found that regardless of the soil type, an optimum degree of compactness for crop growth is 85%–87% of the achieved maximum level (Håkansson, 1990a,b). This procedure is an interesting example how to go around different behaviors of different soil types (Fig. 2.6). The main drawback for forest soils is that our soils are vegetated, contain roots, and the soils are stony and layered, which means we will have great troubles to get undisturbed soil cores. The other thing that Håkansson's studies may teach us is that the soil can get too loose. Below the optimum level of compaction, the capillarity is broken and the plants may have difficulties to get enough of water. This aspect is rather little known about in forestry. But planting experiments on excavated mounds have shown that too loose soil on sandy soils may hazard the plant growth during dry summers (Örlander and Gemmel, 1989). Similarly, compaction of seed spots increased the stocking after direct-seeding of Jack pine (*Pinus banksiana* Lamb.) in conjunction with scarification (Van Damme et al., 1992).

Conclusion: There are some generally accepted rules to judge if the soil conditions are bad, but still microvariation and adaptation to the new situation may alleviate the impact of soil compaction.

Wind-throwing

Quite little is known about the effects of soil compaction and root damage on wind-throwing of trees. Root penetration will be impeded in a compacted soil and thus the anchoring of the trees is less good. This can be seen on very stony soils or shallow soils. The first trees to blow down are the trees situated on those localities. The problem to evaluate the risks of less good anchoring of the trees is the occasionally event we get a heavy storm. If the storms were easily foreseen, such experiments could be easier designed.

In newly thinned stands, the risk of wind-throwing is high because of the openings in the stand subjecting unadapted trees to the wind (Persson, 1975). Root damage along the strip roads will decrease the stability of the trees. Increased rot infections due to root wounds will further reduce the stability of the trees. However, the role of root damage in the total risk for wind-throwing appears not to be proven yet.

Erosion

The extent of erosion may be related to grade, length of slope, intensity of use, soil vegetation, and climatic factors (Trimble and Weitzman, 1953). Two of those factors, extent of soil disturbance and soil vegetation, are influenced by the logging operations.

Clear-felling operations and logging roads are reported to increase sediment concentrations in the run-off (Froehlich, 1978; Johnson and Beschta, 1980; Larsson and Gretener, 1982; McCashion and Rice, 1983). In a study on agricultural land, it was found that compaction from wheel traffic increased the bulk density with 30% and decreased total porosity of the tilled layer, both which reduced the infiltration (50%), increased the run-off (1.5 times), and doubled the amount of soil loss (Young and Voorhees, 1982).

In a compacted soil, the rain water infiltration is severely impeded (Johnson and Beschta, 1980; Hofmann, 1989) which means water will accumulate in the wheel ruts after a heavy rain, and the run-off can start from this position. The run-off can concentrate in or near existing roads, adding to stresses already placed upon the drainage system. When the run-off becomes heavy and rapid enough to carry away soil particles, the erosion starts. Erosion can decrease site productivity due to the loss of fertile soil and the sediment produced can damage fish habitats. However, leaching from a clear-felled area may depend to a great extent on the removal of the trees (cf. Smedegaard Mark and Thorn Clausen, 1993) and soil disturbance from logging machinery may give additional effects. In general, the leaching of nutrients will increase the more the soil is exposed (Lundmark, 1986).

"Obviously, preventing undue soil disturbance will not only conserve site productivity but will also help to minimize downstream problems of water quality" (Johnson and Beschta, 1980).

The other aspect is the presence of live vegetation cover. A burning of the area may increase the sediment flow considerably. The vegetation improves rain water infiltration but also the rooting of the vegetation will keep the soil on place and improve the soil stability (Waldron, 1977; Johnson and Beschta, 1980; O'Loughlin et al., 1982; Watson and O'Loughlin, 1985). The tensile strength of soil is three–five orders of magnitude weaker than that of roots under tension (Coutts, 1983). Soil slip will occur many times more readily when roots are few or absent than when the soil is permeated with roots.

If the stand is growing on compacted soil, the trees could be expected to have a superficial rooting. In vulnerable soils on slopes, a slip zone may develop under the rooting zone and a landslide can occur after heavy rains although the soil was covered by vegetation (cf. Anon, 1981).

Conclusion: Erosion and leaching after badly organized logging operations on slopes are rather well documented and it must be pursued to expose as little uncovered soil as possible. However, the effects of soil compaction on erosion and leaching are less elucidated.

Nutrient situation

Soil compaction and soil disturbance may reduce the nutrient content of the top soil, especially the available P, Ca, and K (Miller and Sirois, 1986) and the microbial immobilization might be slowed down due to poor aeration. Soil disturbance, e.g., due to dragging of logs, may reduce the organic matter (Miller and Sirois, 1986), which affects the moisture holding capacity and amount of available nutrients. On the other hand, a rut formation with breakage of the humus layer and the mixing of the soil and the humus together might be favorable for the microbial turn-over. Application of bark mulch prior to compaction tended to reduce compaction effects and improved growth of *Acer rubrum* L. and *Quercus rubra* L., but not of *Quercus velutina* Lam. (Donnelly and Shane, 1986), which may indicate that the use of putting slash on logging roads can both protect the soil from severe compaction and also improve the growing conditions. Fertilization of compacted soils can improve tree or seedling growth (Hatchell, 1981; Wästerlund, 1985), and it might be interpreted as a poor root penetration in the soil volume is compensated by the fertilization rather than a reduced nutrient content in the soils.

Conclusion: Soil compaction may reduce the availability of nutrients, e.g., due to reduced microbial turn-over as a result of poor aeration, but the evidences are not well established.

Water quality

By opening the soil, there is always an increased risk for leaching from the soil (cf. Lundmark, 1986). Thus, severe ground disturbance at logging operations will always increase the leaching and the risk for erosion, as discussed above. In watershed studies, the nutrient composition in the streamflow has been studied after clear-felling or site preparation activities (cf. Harr et al., 1979). However, the effect of soil compaction is then mixed with other activities. The great increase of, e.g., nitrate concentrations in the run-off water in those cases (Rosén, 1982, 1984) indicates that even a severe soil damage at thinning operations could very well be harmful for the downstream water quality.

Other aspects

Roots are the major sites of synthesis of gibberellins and cytokinins and can therefore no longer be regarded solely as absorbing or anchoring organs (Russell, 1977). Ethylene is perhaps the endogenous growth substance which appears closest associated with the response of plants to anaerobic conditions (Russell, 1977). The production of ethylene is increased by mechanical wounding, flooding, and drought, and it gives effects resembling mechanical impedance on roots (Russell, 1977), which makes it difficult to evaluate the cause and the reason for it. Rather high concentration has been measured in anaerobic soils (10 ppm), whereas 2 ppm or less can restrict root extension in some species (Russell, 1977; Fitter and Hay, 1981), and it may play a major role in the injury of plants in such conditions. The significance of ethylene produced in soil is uncertain and largely unexplored but production of the substance by soil-inhabiting fungi has been shown (Dasilva et al., 1974), and it seems to be stimulated by anaerobic conditions.

Allelophatic substances from root exudates, leaf exudates, or decomposing by-products may inhibit root development. Such allelophatic substances are known from lichens (Brown and Mikola, 1974), heather (*Calluna vulgaris* L., Handley, 1963), crowberry (*Empetrum hermaphroditum* Hagerup, Nilsson and Zackrisson, 1992), and bracken (*Pteridium aquilinum*

(L.) Kuhn. Gliessman, 1976), and they may be considered as a part of a competition weapons different plants exhibit. How soil compaction will influence the release of these substances is not known.

Finally, it should be noticed that even trees will compact the soil themselves. The weight of a tree is transferred down to the soil by the roots. With a flat superficial root system a considerable compaction can be achieved just because of the weight of the tree (Hartge et al., 1983; Ryan and McGarity, 1983). A contributing factor is the tree swing in the wind. Hintikka (1972) showed that Norway spruce trees in the wind moved the roots several millimeters vertically. The most lateral movement was on a radius of about 0.5—1 m from the stem base. The presence of tree roots will therefore modify the soil conditions and contribute to increased variability of the forest soil (Ryan and McGarity, 1983; Vogt et al., 1991).

Thus, very little is actually known about the chemical aspects of soil compaction and soil disturbance.

Conclusions

Concerning the knowledge about the impacts of soil compaction on the sustainable tree growth, I would like as conclusions make two quotations. The first one is from Voorhees (1992):

> *Increasing values of both bulk density and penetrometer resistance are generally considered as detrimental to root growth. This may be universally true if only an isolated individual root is considered. But when a root system is considered, especially under field conditions, factors such as heterogenous soil conditions, compensatory root growth, and time depth-dependent changes in root function make it virtually impossible to generalize a relationship between soil conditions and root growth.*

As the second quotation I take three of the six recommendations to the research challenges for the 1990s by Zobel (1992):

> *1. To develop interdisciplinary teams (including modelers) to study the integration of edaphic and root characteristics (including micro- and macroorganisms) and produce a holistic view of the rhizosphere.*
>
> *2. To document how these plant reponses in turn modify the edaphic characteristics with which they interact.*
>
> *3. To determine which edaphic characteristics (both quantitative and qualitative) plant root systems are actually adapted to and the mechanism(s) of that adaptation.*

References

Ågren, G., Axelsson, B., 1980. Annual carbon budgets for Scots pine. Swed. Univ. Agr. Sci., Dept. of Ecol. and Env. Res., Uppsala. Swed. Con. For. Proj., Techn. Rep. No 25 145–152.

Ågren, G., Axelsson, B., Flower-Ellis, J.G.K., Linder, S., Persson, H., Staaf, H., Troeng, E., 1980. Annual carbon budget for a young Scots pine. In: Persson, T. (Ed.), Structure and Function of Northern Coniferous Forests - an Ecosystem Study. Ecol. Bull. (Stockholm) 32, pp. 307–313.

Anon, 1981. Tree Roots and Slope Stability. Forest Research Institute, Rotorua. What's new in forest research, No. 104.

Bennie, A.T.P., 1991. Growth and mechanical impedance. In: Waisel, Y., Eshel, A., Kafkafi, U. (Eds.), Plant Roots. The Hidden Half. Marcel Dekker, Inc., New York, pp. 393–410.

Bernhardsson, A., Martinsson, L., 1986. Rötfrekvens i högproduktiva granskogar (Root rot frequency in highly productive spruce forests). Swedish University of Agricultural Sciences, Department Forest engineering school, Skinnskatteberg. Diploma work. (Swedish).

Bilan, V.M., 1960. Root Development of Loblolly Pine Seedlings in Modified Environments. Department of Forestry, Stephen F. State College, Texas. Bull. No 4.

Binkley, D., 1986. Forest Nutrition Management. Wiley, New York, 290 p.

Brown, R.T., Mikola, P., 1974. The influence of fruticose soil lichens upon the mycorrhizae and seedling growth of forest trees. Acta For. Fenn. 141, 23 pp.

Burroughs Jr., E.R., Thomas, B.R., 1977. Declining root strength in Douglas-fir after felling as a factor in slope stability. U.S.D.A. For. Serv. Res. Pap. INT-190.

Coutts, M.P., 1983. Root architecture and tree stability. Plant Soil 71, 171–188.

Coutts, M.P., Philipson, J.J., 1978. Tolerance of tree roots to waterlogging. II. Adaptation of Sitka spruce and lodgepole pine to waterlogged soil. New Phytol. 80, 71–77.

Dasilva, E.J., Henriksson, E., Henriksson, L.E., 1974. Ethylene production by fungi. Plant Sci. Lett. 2, 63–66.

Donnelly, J.R., Shane, J.B., 1986. Forest ecosystem responses to artificially induced soil compaction. I. Soil physical properties and tree diameter growth. Can. J. For. Res. 16, 750–754.

Ehlers, W., 1975. Observations on earthworm channels and infiltration on tilled and untilled loess soil. Soil Sci. 119, 242–249.

Fayle, D.C.F., 1980. Secondary thickening in tree roots and environmental influences. In: Sen, D.B. (Ed.), Environment and Root Behavior. Geobios, Jodphur, pp. 93–118.

Fayle, D.C.F., Pierpoint, G., 1975. Evolvement of root and shoot systems of red pine on soils of different moisture availability. In: Bernier, B., Wingert, C.H. (Eds.), Forest Soils and Forest Land Management. Les Presses de lUniversité Laval, Ouebec, pp. 11–25.

Fitter, A.H., Hay, R.K.M., 1981. Environmental Physiology of Plants. Academic press, London.

Forristall, F.F., Gessel, S.P., 1955. Soil properties related to forest cover type and productivity on Lee forest, Snohomish county, Washington. Soil Sci. Soc. Am. Proc. 19, 384–389.

Froehlich, H.A., 1978. The influence of clearcutting and road building activities on landscape stability in western United States. In: Youngberg, C.A. (Ed.), Forest Soils and Land Use. Proc. 5th North American Forest Soils Conf., Colorado State Univ., Ft. Collins, CO, pp. 165–173.

Gaiser, R.N., 1952. Root channels and roots in forest soils. Proc. Soil Sci. Soc. Am. J. 16, 62–65.

Gill, C.J., 1970. The flooding tolerance of woody species: a review. For. Abstr. 31, 671–688.
Gilman, E.F., Leone, I.A., Flower, F.B., 1987. Effect of soil compaction and oxygen content on vertical and horizontal root distribution. J. Environ. Hortic. 5 (1), 33–36.
Gliessman, S.R., 1976. Allelophaty in a broad spectrum of environments as illustrated by bracken. Bot. J. Linn. Soc. 73, 95–104.
Greenwood, D.J., 1969. Effect of oxygen distribution in the soil on plant growth. In: Whittington, W.J. (Ed.), Root Growth. Butterworths, London, pp. 202–221.
Håkansson, I., 1990a. A method for characterizing the state of compactness of the plough layer. Soil Tillage Res. 16, 105–120.
Håkansson, I., 1990b. Soil compaction control - objectives, possibilities and prospects. Soil Technol. 3, 231–239.
Handley, W.R.C., 1963. Mycorrhizal Associations and Calluna Heathland Afforestation. H.M.S.O., Forestry Commision, London. Bulletin No 36.
Harr, R.D., Fredriksen, R.L., Rothacher, J., 1979. Changes in Streamflow Following Timber Harvest in Southwestern Oregon. U.S.D.A., For. Serv. Res. Pap., PNW-249.
Hartge, K.H., Bläsing, D., Herklotz, K., 1983. Veränderungen des Bodengefüges unter dem Einfluss mehrjähriger Baumwurzeln. Forstw. Cbl. 102, 99–110.
Hartmann, F., 1951. Root types as site indicators. In: Der Waldboden: Humus-, Boden-, und Wurzeltypen als Standortsanzeiger. Produktiv.-Zentrum, Wien, pp. 123–152.
Hatchell, G.E., 1981. Site preparation and fertilizer increase pine growth on soils compacted in logging. South. J. Appl. For. 5, 79–83.
Heilman, P., 1981. Root penetration of Douglas-fir seedlings into compacted soil. For. Sci. 27, 660–666.
Hintikka, V., 1972. Wind-induced root movements in trees. Commun. Inst. For. Fenn. 76, 1–56.
Hofmann, R., 1989. Bodenschäden durch Forstmaschineneinsatz. Forstwissenscaftlichen Fakultät der Universität Freiburg i Br. Diss.
Jackson, M.B., Drew, M.C., 1984. Effects of flooding on growth and metabolism of herbaceus plants. In: Kozlowski, T.T. (Ed.), Flooding and Plant Growth. Academic Press, Orlando, Florida, pp. 47–128.
Jakobsen, B.F., Greacen, E.L., 1985. Compaction of sandy forest soils by forwarder operations. Soil Tillage Res. 5, 55–70.
Jaynes, D.B., 1990. Temperature variations effect on field-measured infiltration. Soil Sci. Soc. Am. J. 54, 305–312.
Jenny, H., 1980. The Soil Resource. Springer Verlag, New York.
Johnson, M.G., Beschta, R.L., 1980. Logging, infiltration capacity, and surface erodibility in western Oregon. J. For. 78, 334–337.
Klimo, E., 1983. The influence of clearcut logging on soil properties and the cycle of elements in the ecosystem of spruce forest. Lesnictvi 29 (3), 497–512.
Laitakari, E., 1927. The root system of pine (*Pinus silvestris*). A morphological investigation. Acta For. Fenn., 33.1 (Finnish, English extended summary).
Larsson, S., Gretener, B., 1982. Effekten av skogsavverkning på erosionsförloppet i sedimentsluttningar iövre Klarälvsdalen. Statens Naturvårdsverk, Solna. SNV PM 1601. (Swedish).
Lenhard, R.J., 1986. Changes in void distribution and volume during compaction of forest soil. Soil Sci. Soc. Am. J. 50, 462–464.
Lohm, U., 1977. Samspelmellanmarkorganismer. Sveriges Skogsvårdsförb. Tdskr. 75 (2–3), 141–147 (Swedish).
Lohm, U., Lundkvist, H., Persson, T., Wirén, A., 1977. Effects of Nitrogen Fertilization on the Abundance of Enchytraeids and Microarthropods in Scots Pine Forests. Studia Forestalia Suecica No 140.

Lundmark, J.-E., 1986. Skogsmarkens ekologi. [Forest Floor Ecology]. Del 1. Grunder. Skogsstyrelsen. (Swedish), Jönköping.

McCashion, J.D., Rice, R.M., 1983. Erosion on logging roads in northwestern California: how much is avoidable? J. For. 81, 23–26.

Meyer, F.H., 1967. Feinwurzelverteilung bei Waldbaumen in Abhängigkeit vom Substrat. Forstarchiv 38, 286–290.

Miller, J.H., Sirois, D.L., 1986. Soil disturbance by skyline yarding vs. skidding in a loamy hill forest. Soil Sci. Soc. Am. J. 50, 1579–1583.

Minchin, F.R., Pate, J.S., 1973. The carbon balance of a legume and the functional economy of its root nodules. J. Exp. Bot. 24, 259–271.

Minore, D., Smith, C.E., Woolard, R.F., 1969. Effects of high soil density on seedling root growth of seven Northwestern tree species. U.S.D.A., For. Serv., Pacific Northwest For. Range Exp. Stn., Portland. Res. Note PNW-112.

Nambiar, E.K.S., Sands, R., 1992. Effects of compaction and simulated root channels in the subsoil on root development, water uptake and growth of radiata pine. Tree Physiol. 10, 297–306.

Nilsson, M.-C., Zackrisson, O., 1992. Inhibition of Scots pine seedling establishment by *Empetrum hermaphroditum*. J. Chem. Ecol. 18, 1857–1870.

Olsson, M.T., 1986. Micromorphometric evaluation of artificial compaction of fine sandy till. For. Ecol. Manag. 17, 109–117.

Örlander, G., Gemmel, P., 1989. Markberedning. [soil preparation]. Sveriges Skogsvårds-förb. Tdskr. 3–89, 53 (Swedish).

O'Loughlin, C.L., Rowe, L.K., Pearce, A.J., 1982. Exceptional storm influences on slope erosion and sediment yield in small catchments, North Westland, New Zealand. In: First National Symposium on Forest Hydrology. Australia Institution of Engineers, National Conference Publication No 82/6, pp. 84–91.

Persson, P., 1975. Windthrow in Forests - It's Causes and the Effect of Forestry Measures. Royal College of Forestry, Department of Forest Yield Research, Stockholm. Res. Note 36. (Swedish, English summary).

Persson, H., 1980. Death and replacement of fine roots in a mature Scots pine stand. In: Persson, T. (Ed.), Structure and Function of Northern Coniferous Forests - an Ecosystem Study. Ecol. Bull, pp. 251–260 (Stockholm) 32.

Persson, J., Nasholm, T., 2018. Greater carbon allocation to mycorrhizal fungi reduces tree nitrogen uptake in a boreal forest. Ecology 97 (4), 1012–1022.

Rastin, N., 1991. Influence of waterlogging on root distribution, fine root biomass and mycorrhizal number of Norway spruce. In: McMichael, B.L., Persson, H. (Eds.), Plant Roots and Their Environment. Proc. Of an ISRR-Symposium, August 21-26, 1988, Uppsala, Sweden. Developments in Agriculture and Managed-Forest Ecology 24. Elsevier, Amsterdam, pp. 319–332.

Romell, L.-G., 1939. Barrskogensmarksvampar och deras roll i skogensliv. Sv. Skogs-vårdsförb. Tidskr. 37, 348–375 (Swedish, English summary).

Rosén, K., 1982. Supply, Loss and Distribution of Nutrients in Three Coniferous Forest Watersheds in Central Sweden. Swedish University of Agricultural Sciences, Department of Forest Soils, Uppsala. Report No 41.

Rosén, K., 1984. Effect of clear-felling on runoff in two small watersheds in central Sweden. For. Ecol. Manag. 9, 267–281.

Russell, R.S., 1977. Plant Root Systems. Their Function and interaction with the Soil. McGraw-Hill Book Company Limited, London.

Ryan, P.J., McGarity, J.W., 1983. The nature and spatial variability of soil properties adjacent to large forest eucalypts. Soil Sci. Soc. Am. J. 47, 286–293.

Sands, R., Greacen, E.L., Gerard, C.J., 1979. Compaction of sandy soils in radiata pine forests. I. A penetrometer study. Aust. J. Soil Res. 17, 101–113.

Schalkwijk-Barendsen, H.M.E., 1991. Mushrooms of Northwest North America. Lone Pine Publishing Auburn WA.

Scherfose, V., 1990. Feinwurzelverteilung und Mykorrhizatypen von Pinus sylvestris in verschiedenen Bodentypen. Universität Göttingen. Berichte des Forschungszentrums Waldökosysteme, Reihe A, Bd, 62. Diss.

Schulze, E.D., Fuchs, M.I., Fuchs, M., 1977. Spatial distribution and photosynthetic capacity and performance in a mountain spruce forest of northern Germany. Oecologia (Berl.) 29, 43–61.

Schuster, C.E., 1936. Root development of trees as affected by physical properties of the soils. In: Proc. 32nd Ann. Meet., Washington State Hort, pp. 22–26.

Sharpley, A.N., Meisinger, J.J., Power, J.F., Suarez, D.L., 1992. Root extraction of nutrients associated with long-term soil management. In: Hatfield, J.L., Stewart, B.A. (Eds.), Limitations to Plant Root Growth. Advances in Soil Science, vol. 19. Springer-Verlag, New York, pp. 27–52.

Simmons, G.L., Pope, P.E., 1987. Influence of soil compaction and vesicular-arbuscular mycorrhizae on root growth of yellow poplar and sweet gum seedlings. Can. J. For. Res. 17, 970–975.

Smedegaard Mark, H., Thorn Clausen, J., 1993. Reforesting of a Sludge Applicated Conifer Plantation in Hesselvig. Hedeselskabets Forskningsvirksomhed. Beretning nr 54 (Danish, English summary).

Smeltzer, D.L.K., Bergdahl, D.R., Donnelly, J.R., 1986. Forest ecosystem responses to artificially induced soil compaction. II. Selected soil microorganism populations. Can. J. For. Res. 16, 870–872.

Stolzy, L.H., Focht, D.D., Flühler, H., 1981. Indicators of soil aeration status. Flora 171, 236–265.

Sutton, R.F., 1991. Soil Properties and Root Development in Forest Trees: A Review. Forestry Canada, Ontario Region. Information Report O-X-413.

Tamm, C.O., 1974. Experiments to analyse the behavior of young spruce forest at different nutrient levels. In: Cavé, A.J. (Ed.), Structure, Functioning and Management of Ecosystems. Proc. 1st Int. Congr. Ecol. Pudoc, Wageningen, pp. 266–272.

Taylor, H.M., Klepper, B., 1978. The role of rooting characteristics in the supply of water to plants. Adv. Agron. 30, 99–128.

Trimble Jr., G.R., Weitzman, S., 1953. Soil erosion on logging roads. Soil Sci. Soc. Am. Proc. 17, 152–154.

Van Damme, L.W., Buse, L., Warrington, S., 1992. Microsite soil compaction enhances establishment of direct-seeded Jack pine in Northwestern Ontario. North. J. Appl. For. 9 (3), 107–112.

Vater, H., 1927. Die Bewurzelung der Kiefer, Fichte und Buche. Tharandter Forstl. Jahrb. 78, 65–85.

Vogt, K.A., Vogt, D.J., Bloomfield, J., 1991. Input of organic matter to the soil by tree roots. In: McMichael, B.L., Persson, H. (Eds.), Plant Roots and Their Environment. Proc. Of an ISRR-Symposium, August 21-26, 1988, Uppsala, Sweden. Developments in Agriculture and Managed-Forest Ecology 24. Elsevier, Amsterdam, pp. 171–190.

Vomocil, J.A., 1965. Porosity. In: Black, C.A. (Ed.), Methods of Soil Analysis, Part I. ASA Monograph No 9. American Society of Agronomy, Madison, WI, pp. 299–314.

Voorhees, W.B., 1992. Wheel-induced soil physical limitations to root growth. In: Hatfield, J.L., Stewart, B.A. (Eds.), Limitations to Plant Root Growth. Advances in Soil Science, vol. 19. Springer-Verlag, New York, pp. 73–96.

Voronkov, N.A., 1974. Root systems of *Pinus silvestris* L. on sandy soils. In: Hoffman, G. (Ed.), Ökologie und Physiologie des Wurzelwachstums. II. Int. Symp. Akademie-Verlag, Berlin, pp. 155–160.

Waldron, L.J., 1977. The shear resistance of root-permeated homogenous and stratified soil. Soil Sci. Soc. Am. J. 41, 843–849.
Wästerlund, I., 1983. Growth reduction of trees near the strip road resulting from soil compaction and damaged roots. - a literture review. Sveriges Skogsvårdsförb. Tdskr. 81 (2), 97–109 (Swedish, English summary).
Wästerlund, I., 1985. Compaction of till soils and growth tests with Norway spruce and Scots pine. For. Ecol. Manag. 11, 171–189.
Wästerlund, I., 1989. How is the occurrence of mushroom fruit bodies influenced by the silvicultural treatments? Svensk Bot. Tidskr. 83, 103–112 (Swedish, English summary).
Watling, R., 1973. Identification of the Larger Fungi. Hulton Group Keys, Amersham.
Watson, A., O'Loughlin, C., 1985. Morphology, strength, and biomass of manuka roots and their influence on slope stability. N. Z. J. For. Sci. 15, 337–348.
Wilcox, H., 1955. Regeneration of injured root systems in noble fir. Bot. Gaz. 116, 221–234.
Woolridge, D.D., 1970. Chemical and physical properties of forest litter layers in central Washington. In: Youngberg, C.T., Davey, C.B. (Eds.), Tree Growth and Forest Soils. Oregon State Univ. Press, Corvallis, pp. 327–337.
Yeager, A.F., 1935. Root systems of certain trees and shrubs grown on prairie soils. J. Agric. Res. 51, 1085–1092.
Young, R.A., Voorhees, W.B., 1982. Soil erosion and runoff from planting to canopy development as influenced by tractor wheel-traffic. Trans. ASAE 25, 708–712.
Zobel, R.W., 1992. Soil environment constraints to root growth. In: Hatfield, J.L., Stewart, B.A. (Eds.), Limitations to Plant Root Growth. Advances in Soil Science, vol. 19. Springer-Verlag, New York, pp. 27–52.

CHAPTER 3

Impeded root growth in compacted till soils

Introduction

This study was originally published by Wästerlund, 1985.

Old root studies started with Aaltonen (1923); Vater (1927) and Laitakari (1927, 1935), and they dug out the whole root system from both birch, pine, and spruce trees and some long laterals were measured to be more than 20 m long.

The first roots to appear from a seed are called primary roots and when they branch first the root hairs are growing out from the epidermis cells and disappear (degenerate) when the root is getting older and a cortex layer is formed under epidermis. Lateral roots are formed from the pericycle and form first long laterals (very long roots) and some of the roots grow in thickness. Fine root system (<2.2 mm in diameter) is well developed in the superficial part of the soil for most of the tree species (Köstler et al., 1968). The vertical pressure (ground pressure) and shearing forces beneath a contact area of a terrain vehicle wheel of a terrain vehicle wheel may damage and break off superficial roots (fine roots), but the trees may also compress the soil by their own weight (a big tree of 10 $m^3 \times 0.9$ $m^3 = 9$ tonne green weight).

Growth reduction may be caused either by direct mechanical injuries to the roots or by hindrance of root growth and change of the root environment by soil compaction. However, these effects probably interact causing a more severe growth reduction than if they had acted individually (Bredberg and Wästerlund, 1983). In agriculture, the effects of soil compaction have been investigated especially within humid and temperate climate zones such as North Europe and North America (Eriksson, 1982; Soane et al., 1982).

The proportion of macropores (>30 μm) diminishes as the soil becomes compressed (Dickerson, 1976; Eriksson, 1982) The orientation of macropores also changes, becoming more horizontal (Ingelög et al., 1977), which decreases the infiltration rate to a greater extent than would be expected by

Soil and Root Damage in Forestry
ISBN 978-0-12-822070-2
https://doi.org/10.1016/B978-0-12-822070-2.00003-1
© 2020 Elsevier Inc.
All rights reserved.

the reduction of larger pores (Steinbrenner and Gessel, 1955; Dickerson, 1976; Johnson and Beschta, 1980). This increases the risk of erosion in logging areas (cf. Trimble and Weitzman, 1953; Young and Voorhees, 1982; McCashion and Rice, 1983). In soils with high bulk densities, the root growth is restricted by mechanical impedance, poor aeration, and poor water availability (Eavis, 1972; Hildebrand, 1983). In North America, soil compaction in forest areas is said to cause up to 30% reduction in growth (Froehlich, 1976; Adams and Froehlich, 1981).

Perry (1964) showed that loblolly pine plantation in old ruts produced only 46% as much as trees planted in surrounding field. The old road was still compacted, although it had not been used for 26 years. On strip roads used occasionally, such as primary and secondary trails, the compaction should not be of that magnitude. Nevertheless, fairly severe soil compaction was found to occur even after a few trips with a low-ground pressure skidder (Hatchell et al., 1970; Hatchell and Ralston, 1971; Campbell et al., 1973; Dickerson, 1976; Froehlich, 1978; Hassan, 1978; Johnson et al., 1979; Gent et al., 1983). With an increasing number of trips, the soil density increases but most of the compaction occurs on the first trip (Miles et al., 1981; Raghavan et al., 1981; Sidle and Drlica, 1981).

The easiest way of examining the relationship between soil compaction and growth seems to be to grow seedlings in compacted soil columns where the environment can be controlled (Eriksson et al., 1974). It has thus been shown that seedling and root growth are retarded differently at a given bulk density, depending on soil type and tree species (Minore et al., 1969; Sands and Bowen, 1978; Zisa et al., 1980; Heilman, 1981; Hildebrand, 1983).

In forest areas in Sweden, the soil is usually covered by a 3—10 cm thick humus layer (Hägglund and Svensson, 1982) with brittle-elastic properties, which will moderate the deformation of the soil until it is torn off. The composition of the humus layer varies within small distances, and its mechanical properties are largely determined by the number of tree roots in the layer (Scholander, 1973).

The aim of this chapter was to examine the compressibility of some till soils common in Sweden and to refer the growth of seedlings of Norway spruce and Scots pine to the soil compaction. Because of the complexity of the humus layer and problems in elucidating the compression behavior of till soils, this layer was excluded in the experiments. However, the ground pressure from a vehicle will compact the soil below the humus layer. When planting a clear-cut area, the plant pots will be put into the A2- and B-horizon. Thus, the results of the study have practical implications as well.

Materials and methods

Five of the most common soils (in particle size distribution) in forest areas in Sweden were examined, viz., coarse sand, fine sand, coarse sandy till, medium sandy till, and fine sandy till soil (Hägglund and Svensson, 1982). A silt loam soil was also subjected to some experiments. The particle size distribution of each soil is shown in Fig. 3.1 and some physical and chemical properties are given in Table 3.1.

All soils were obtained near Garpenberg in central Sweden. The geology of Garpenberg is very variable and several soil investigations have been made in this area (e.g., Lundqvist and Hjelmqvist, 1941; Du Rietz, 1968; Lyford and Troedsson, 1973). The till soils are all podzols developed on a stony, sandy glacial moraine (till) and have numerous stones on the surface and throughout the solum.

The fine sandy till is derived from slightly reddish leptite rock. The soil is above the postglacial shoreline (this soil is described by Lyford and Troedsson, 1973). The medium sandy till is mostly derived from red granite rock. The coarse sandy till is also mostly derived from red granite rock, but is below the postglacial shoreline and is therefore surged. The coarse and fine sands and the silt loam are all sediments (alluvium) found in depressions in the terrain below the postglacial shoreline.

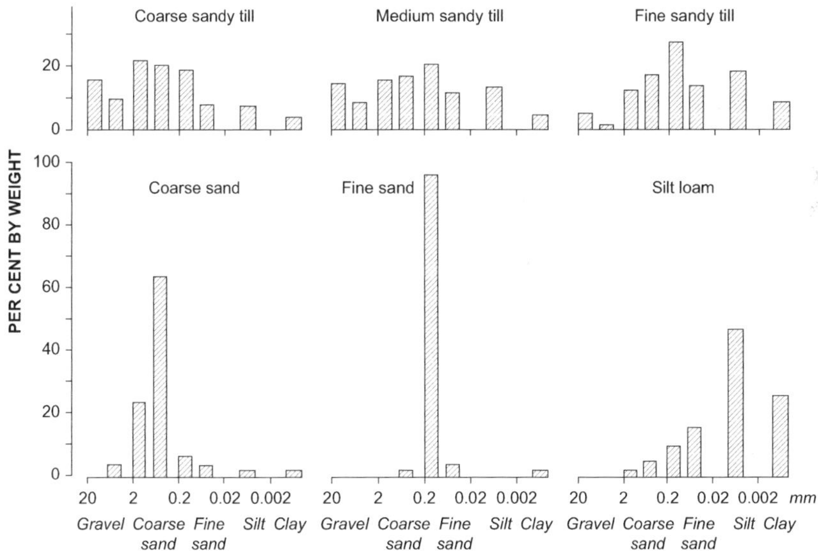

Figure 3.1 Particle size distribution of tested soils.

Table 3.1 Physical properties of tested soils and nutrient content in the till soils.

C	Coarse sandy till	Medium sandy till	Fine sandy till	Coarse sand	Fine sand	Silt loam
Natural bulk density (g cm^{-3})	0.95	0.95	1.25	1.30	1.25	1.40
Specific gravity (g cm^{-3})	2.65	2.60	2.65	2.66	2.67	2.62
Pore volume[a]	64	65	53	51	53	47
Organic matter	4.0	4.7	1.8	1.1	0.3	1.01
$N-NH_4$ mg/100 g	0.89	1.4	0.50	nd	nd	nd
N NO_3 mg/100 g	0.35	1.4	0.19	nd	nd	nd
N tot mg/100 g	1100	1400	340	nd	nd	nd
P−Al mg/100 g	0.37	1.05	1.45	nd	nd	nd
K−Al mg/100 g	1.0	6.0	2.9	nd	nd	nd
Mg− Al mg/100 g	58	3.3	36			

[a]Pore volume $= 1 - \left(\frac{\text{bulk density}}{\text{spec gravity}}\right) \times 100$

Specific gravity was obtained by using alcohol in the pycnometer. Pore volume was calculated based on bulk density (dry weight of soil sample/actual volume) and specific gravity. Organic matter was determined at +600°C for 1 h and the value was corrected according to Ekström (1927).

NH_4−N and $N0_3$−N were determined colorimetrically after soaking the samples in KCl solution. The latter solution was also passed through a cadmium reductor. Total nitrogen content was determined by the Kjeldahl method and available phosphorus, potassium, and magnesium were determined by the ammonium lactate solution method (Egner et al., 1960).

Compaction

The tested soils were collected from the upper part of the B-horizon in forest areas. The amount of stones and gravel made it impossible to take out intact soil cores directly from the till soils. Therefore, all soils were packed by hand, except for the few cores that could be taken for determination of the natural bulk density. Gravel particles coarser than 15 mm were removed to prevent difficulties when packing the soil cores. Other particle sizes were kept giving the soil a more or less natural behavior.

The soil was scooped into aluminum tubes 69 mm in diameter and 70 mm long, and with a volume of 262 cm^3. Other tube lengths were also tested. The inside of the cylinder wall in the tubes used was polished before refilling. When packing the soil cores the soil was tapped with a spoon to prevent formation of air pockets in the soil.

Knowing that approximately 270 g of dry till soil were needed to fill one cylinder, it was easy to weigh the same amount of soil in to each cylinder and to calculate the amount of water required to moisten the soil if the actual moisture content was lower than 20%.

After wetting, if necessary, the soil was mixed and allowed to recover its equilibrium for some weeks in aluminum cans in plastic bags. If larger quantities of water had to be added, the water was given in small doses at intervals of 1 or 2 days. After wetting, the dry fine sandy till soil had to be stored for at least 3–4 weeks.

The soil cores were loaded using a pneumatic cylinder powered by adjustable air pressure. The piston rod acted on a piston fitting into the soil cylinder with a rubber washer on the end that prevented soil particles from coming between the cylinder wall and the piston. Excessive water was allowed to drain out in slits between the soil core and bed plate. Soil cores were individually loaded with 0, 100, 200, or 400 kPa for 10 min when the soil cores were 70 mm high, and for 15 min when the soil cores were 100 or 140 mm high. Duplicates of soil cores were taken randomly in each experiment for determination and control of moisture content on a dry weight basis and as a control of the dry weight (105°C in 24 h). Measurements of the actual height of each soil core before and after loading gave the volume of the soil core before and after loading.

The fine sandy till at 0, 100, and 400 kPa load treatments was subjected to laboratory determinations of distribution of pore sizes by micromorphological analysis (see Olsson, 1985).

Growth of seedlings

After compaction, the soil cores were sown with either Norway spruce (*Picea abies* (L.) Karst.)or Scots pine (*Pinussylvestris* L.). The controlled provenance seed had a germinability of 85%–95%. Ten seeds were sown in a circle in the middle of the top of the soil column, thus minimizing the number of roots searching out around the cylinder wall.

Tubes with seeds were placed on a 30-mm deep bed of sand in a miniature plastic greenhouse in the laboratory, and kept in light 16 h at 20–24°C. After sowing, the seeds were sprinkled with water and then watered about twice to maintain a moisture content of 15%–20%. When the compaction treatments were applied at lower moisture contents, each soil core was watered up to about 25% moisture content.

After 16–20 days, the lengths of the primary roots—no secondary roots were present at that time—in each soil core were recorded after carefully washing the soil from the roots. The radicles were kept in water before measuring to prevent shrinking of the roots. The average root lengths per soil core of well-germinated and well-penetrated seedlings were calculated.

In other test series running for a longer time, the total length of the root system was measured; hence, the lengths of secondary roots, tertiary roots, etc., were also included. These cores were watered with about 10 mL each third day.

The fertilization series were supplied with a full-nutrient solution (Ingestad, 1979; N:P:K = 100:13:65) every third day. The doses were given to correspond to a calculated daily growth rate of 6% (Norway spruce) or 7% (Scots pine), and were based upon the nitrogen content in the nutrient solution and an approximate nitrogen content in the seedling of 0.6% of the fresh weight.

Resulting soil compaction

The range in bulk density between different soil cores with the same soil type, moisture content, and loading was found to be maximally 0.08 g cm^{-3} (95% confidence interval = ±0.06), indicating that the method and the equipment described can reproduce cores of closely controlled bulk density. Variation in moisture content could cause a much larger deviation, as found in previous experiments.

In the comparison of compaction at certain loads of different soil types, the moisture content was set at 20%–25% of the dry weight of the soil. Results of the comparison are shown in Fig. 3.2 and Table 3.2. Increases in bulk density at increasing loads were slight for sand sediments, but considerable for the till soils.

The fine sandy till soil was packed to a bulk density of 1.63–1.75 g cm^{-3} at 400 kPa load. Thus, within the limits of tested soils the finer material there is in a till soil, the easier it will compact. Nevertheless, these till soils compact much more than sediments of the same main particle class even though they have a higher organic matter content than the sediments. The vulnerability of compaction in till soils is most pronounced at slight loadings. This might be due to a loose packing of the soil material, especially as the bulk density of unloaded coarse sandy till soils hardly exceeded 0.95 g cm^{-3} However, it was

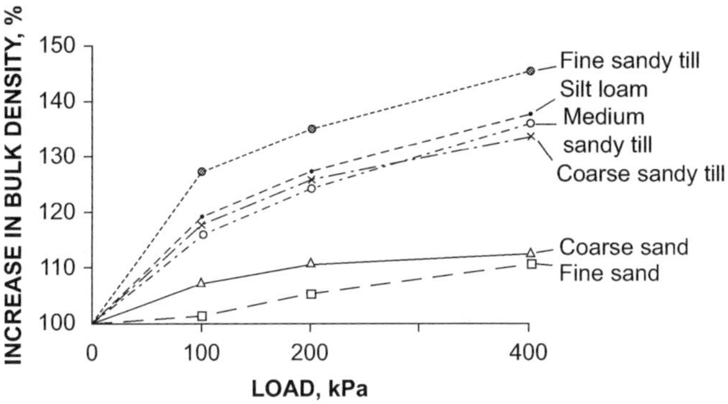

Figure 3.2 Increase in bulk density at different loads, as percentage of control. Moisture content 20%–25%; $n = 6–8$ for till soils, $n = 2–4$ for sediments. Soil cores 70 mm high and 69 mm in diameter. Static load for 10 min.

Table 3.2 Average bulk densities in soil cores after compression.

	Load (kPa)			
Soil	**0**	**100**	**400**	**(n)**
Coarse sandy till	0.98	1.16	1.30	8
Medium sandy till	0.96	1.11	1.30	6
Fine sandy till	1.17	1.49	1.69	8
Coarse sand	1.27	1.37	1.43	4
Fine sand	1.21	1.28	1.36	4
Silt loam	1.12	1.33	1.53	2

quite difficult to place that amount of medium sandy till soil in the compaction tube without loading. When wetted, the dry till soils swelled, thus lowering the bulk density. If the fine sandy till soil was wetted too much, the aggregate structure broke down and the soil attained viscous properties. Thus, the compaction vulnerability of the till soils does not seem to depend on packing that is too loose in the experiments.

Tests of the compression method

According to Dexter and Tanner (1974), the time dependence of compression may very between different soil types. The obtained compression of sand soils was in the same order as obtained by Sands et al. (1979). However, the time dependence of till soils was most interesting.

Being in the middle texture class, the medium sandy till soil was chosen for this experiment. When loaded at 100 kPa the soil compressed quite slowly and reached almost the final bulk density after 5 min (Fig. 3.3). An increase in compression time from 10 min to 1 h resulted in the bulk density increasing by 0.01 g cm^{-3}. At 400 kPa load the soil compressed quickly. Final bulk density increased only 0.07 g cm^{3} between 8 s and 1 h. Thus, the chosen compression time of 10 min seems to correspond well with a final equilibrium of static compression.

As pointed out by Koolen (1974), readings of a uniaxial compression test may be erratic if the friction between soil and cylinder wall induces interfering shear stresses which cause an unevenly distributed loading. If this is the case, the normal stress at the top of the sample is higher than at the bottom, giving compression values that are lower than they should be.

Compression tests were performed on the medium sandy till soil in soil cores with varying heights (35–140 mm), giving diameter/height ratios ranging from 0.5 to 2.0. If the influence of wall friction is negligible, the compaction of the soil should be equal in all cases. As shown in Table 3.3, the wall friction does not seem to seriously affect the compression of till soils. This was also proved by measuring the penetration resistance with a cone penetrometer on the top and at the bottom of the compressed till soil

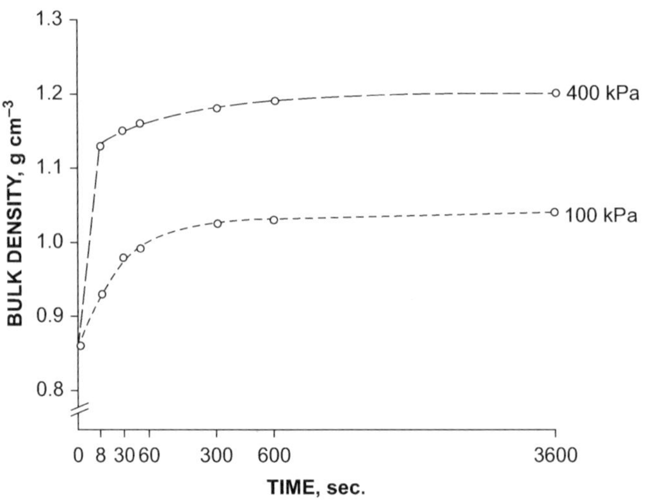

Figure 3.3 Compaction with time <√t) of a medium sandy till soil at two loads, 100 and 400 kPa. Moisture content 25%; soil cores 100 mm high and 69 mm in diameter.

Table 3.3 Soil compaction in percent of control in soil cores with varying heights (*h*) from 35 to 140 mm but the same diameter (D), 69 mm.

D/H	Load (kPa)			
Load	**100**	**200**	**400**	***n***
2.0	119		131	1
1.0	116	124	135	4
0.7	119		138	1
0.5	114		134	2

Coarse to fine sandy till soil loaded for 10 min ($D/h = 1.0-2.0$) and 15 min ($D/h = 0.5-0.7$).
n = number of samples.

sample. The 13 mm cone gave no diverging resistance values either in the middle or at the sides.

Interfering shear stresses may occur with a friction soil such as dry sand and with cohesive soils such as clay. The soils tested here may be classed as brittle soils (McIntyre, 1973) with the exception for the silt loam soil. Compression of the latter soil may be underestimated due to wall friction. Unsaturated till soils seem to have very few elastic properties. When unloading after compression for 10 min no relaxation was visible.

Root growth

The primary roots of pine seedlings grew through the 70 mm tall uncompacted soil core in 16–20 days, whereas spruce roots needed about 25 days.

The variation in root lengths within a single soil core was usually within ±7 mm at a confidence interval of 95%. According to Student's *t-test* an average difference in root lengths between two soil cores, e.g., a control and a compressed one, is usually significant at the $P < .05$ level if the deviation is larger than 15%. Commonly, the variation in root lengths within a soil core increased somewhat with increased compaction. This was most obvious for till soils, where probably the irregularities in the soil affected root growth.

The results show that the more the soil was compacted, the less root growth occurred. Even in the sand sediments the high loading caused a compaction which significantly impeded root growth (e.g., spruce at 400 kPa $P < .01$ and $P < .001$ if the average is taken of two compression series). Presented as root length in percent of the control, the root lengths in different soil types are almost inverse to the compaction curves (Figs. 3.2

and 3.4), i.e., although compressed to lower bulk density, the roots penetrate the coarse sandy till soil much slower than in the coarse sand soil. On the other hand, root growth was more restricted in medium sandy till soil than in the coarse sandy till soil, though the soils were compacted to about the same degree, both in absolute and in relative figures. Thus, the bulk density alone is a poor indicator of the damage of soil compaction unless the behavior of the soil is known. Moreover, the results show that the till soils, which constitute about 50% of forest areas in Scandinavia, compact easily and, what is worse, the compaction might impede root growth much more than the increase in bulk density indicates.

Comparison of the two tree species tested (Fig. 3.4A and B) shows that different tree species are restricted to different degrees of soil compaction. Here, Norway spruce has a much more impeded root growth than Scots pine. The difference in impedance seems to increase with increasing compaction.

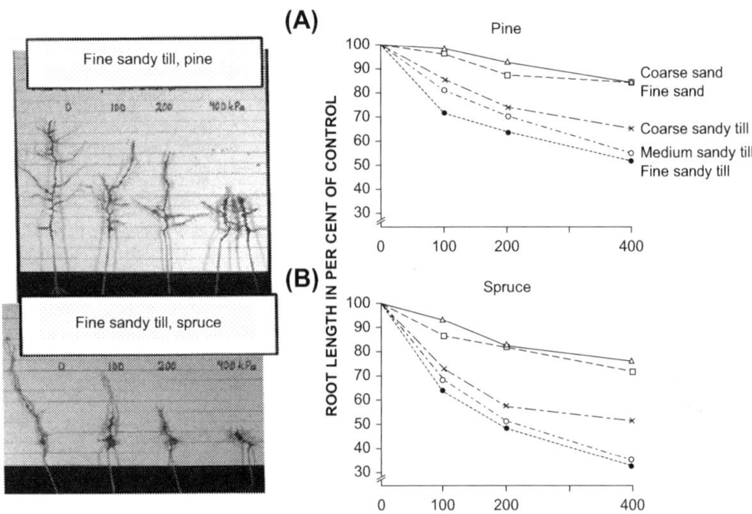

Figure 3.4 Root lengths about 20 days after sowing, as percentage of control, at different levels of soil compaction. Moisture content in the soil at loading 20%–26%. Soil cores 70 mm high and 69 mm in diameter; $n = 2–5$ for till soils and $n = 2$ for sediments with 5–10 well-developed seedlings in each soil core. *(Photo I. Wästerlund, to show the roots from sandy till to illustrate the drawing.)*

Test of method

The root growth results obtained in compacted soil cores may be false due to the short cultivation time. The primary roots might be restricted more than secondary roots. The impedance might also be fictitious due to a delay in the root penetration of the topsoil. The latter argument is weakened by the fact that only well-developed seedlings and roots were measured. Often up to half of the spruce seedlings on compacted till cores were not measured due to poor root penetration of the topsoil.

Growth of tree seedlings into compacted soil cores has not been a widely used method. In the few studies reported, the growing time has been between 35 and 160 days (Sands and Bowen, 1978; Zisa et al., 1980; Heilman, 1981; Hildebrand, 1983). Here, three different additional time intervals, 26, 33, and 120 days, were chosen to study the root growth in compacted soil cores.

In a fine sandy till soil, the seedlings were allowed to grow for an additional 6 or 13 days beyond the ordinary growth time of 20 days. The few roots extending below the soil core into the sand bed were not measured. It was found that the relative root lengths of pine seedlings decreased with time in compacted soil cores, but increased at the end of the period in the case of spruce seedlings (Table 3.4).

In a medium sandy till soil, the seedlings were grown for 120 days. Unfortunately, the day length (16 h) was too short to prevent this provenance of pines from forming top buds. Therefore, the height growth stopped and the effect of soil compaction on top growth could not be measured. The bud formation is regulated by the day length, but root growth continues as long as the temperature is favorable (Dormling, 1975, 1979; Dormling et al., 1977). Spruce seedlings continued to grow in height, but died on the most compacted (400 kPa) soil cores. A load of 100 kPa on the soil here resulted in top growth decreasing to 54% of the top growth in uncompacted soil cores, measured as fresh weight above the top of the soil. Both pine and spruce seedlings had shorter relative root lengths in compacted soil cores after 120 days than after 20 days. The decrease was more pronounced at low loadings.

The results show that the growth of the primary root is impeded by soil compaction. The impeded growth of the primary root seems to induce a luxuriant formation of secondary roots (cf. Fig. 3.3 in Bredberg and Wästerlund, 1983). This was the case after 26 days for pine and 33 days for spruce, resulting in an apparent recovery from the soil compaction. Then, at

Table 3.4 Comparison between measured root growth in compacted soils after 18–20 days and root growth after longer time intervals from sowing.

	Pine				Spruce				
	Root length in control	Root length (% of control)			Root length in control	Root length (% of control)			
Soil type and time period	(mm)	100 kPa	200 kPa	400 kPa	(mm)	100 kPa	200 kPa	400 kPa	No. of series
***Fine sandy till*[a]**									
18–20 days	55	72	64	52	35	65	49	34	3–4
26 days	203	63	58	54	65	58	42	33	1
33 days	435	55	54	40	119	78	66	60	1
***Medium sandy till*[b]**									
18–20 days	62	81	70	61	45	67	52	36	2–3
120 days	1440	66	–	56	680	62	–	–	2
120 days + fertilization	2270	–	–	50	948	–	–	–	1

One trial with 0 and 400 kPa compaction was fertilized with a full-nutrient solution every third day corresponding to a calculated growth rate of 7% (pine) and 6% (spruce), respectively.
[a]Soil cores 70 mm high, 69 mm in diameter; 10 seeds per soil core.
[b]Soil cores 140 mm high, 69 mm in diameter; 6 seeds per soil core.

the extension of secondary roots soil compaction seems to impede their growth even more than the primary root (cf. Taylor, 1974; Tworkoski et al., 1983).

A third trial with medium sandy till soil was fertilized with a full-nutrient solution according to Ingestad (1979). Only uncompacted and heavily compacted cores were used in this experiment. Harvested after 120 days it was found that the pine seedlings in the control had grown at a daily growth rate of about 6.5% and the spruce seedlings about 5.5%. Thus, the calculated growth rate of 7% and 6%, respectively, was only a slight overestimation.

The unfertilized pine seedlings grew at a growth rate of about 5%. This means that the mineral content in the soil cores was a limiting factor for growth during longer periods (cf. Table 3.1). Impeded root growth, i.e., decreased uptake of slowly mobile nutrients (e.g., phosphorus), might be compensated by the fertilization.

Although shoot growth in compacted and fertilized soil cores was more than in compacted and unfertilized soil cores, the shoots were lighter than in fertilized and uncompacted soil cores. Spruce seedlings in compacted soil cores were only 59% by weight of the uncompacted and fertilized, and equal to the unfertilized control.

The root systems of spruce on the heavily loaded soil core were very short and thick, lying on the top of the soil core. This means that the reduction in root length was caused by soil compaction and not nutrient deficiency (cf. Davies and Runge, 1969).

Thus, the method of measuring growth reduction of primary roots is a relatively quick way of estimating the impact of soil compaction on tree seedlings, and merely underestimates rather than overestimates the impact. The results show that even a light static loading of a till soil may severely impede root growth, which also seems to affect the top growth of almost the same magnitude. In poor sites the impact might be greater than in good sites because growth in compacted soil can be increased by fertilization (Hatchell, 1981).

Discussion

Deformation of the soil in the cores occurred mainly with the expulsion of air, whereby the deformation is then in the form of compaction and not consolidation (Dexter and Tanner, 1974). The compaction is dependent on particle size distribution, surface roughness of particles, moisture content, and organic matter.

If the particle size distribution of a soil is such that finer particles can fit between larger particles to cause very small pores, a very high compaction may occur (Veihmeyer and Hendrickson, 1948; Bodman and Constantin, 1965; Gupta and Larson, 1979; Rawls, 1983). A combination of sand with finer material such as silt and clay could thus be strongly and more rapidly compacted. Materials with rougher particle surfaces, such as till soils, have a greater soil particle packing arrangement stability which decreases the elastic properties (Cruse et al., 1980). On the other hand, in an uncompressed condition, these factors, together with frictional resistance, contribute to a high pore volume in the soil (Bodman and Constantin, 1965; Cruse et al., 1980).

Increasing the moisture content up to field capacity will usually increase the compressibility (Bodman and Constantin, 1965; Bertilsson, 1969; Sands et al., 1979). The moisture content chosen in the present study is less than, but in most cases fairly close to, the field capacity. This means that at least the till soils may be compacted more if the moisture content increases to field capacity.

Increased organic matter content decreases soil compactness (Bertilsson, 1969; Sands et al., 1979; Alexander, 1980). In the present study with 2%–4.5% organic matter in till soils (Table 3.1), this means that if the organic matter decreases, the compaction of till soils will increase (cf. Adams, 1973). On the other hand, remoulded soils compress more rapidly than undisturbed soils (Dexter and Tanner, 1974). In the present study all soil types were remoulded before packing. Thus, the till soils in forest areas might be less susceptible to compaction than indicated here. However, sandy loam soil is known to compact strongly even at light loading (Froehlich, 1978), and its particle size distribution is almost comparable with the fine sandy till soil used in the present experiments.

Many studies deal with maximum bulk density for root penetration (e.g.,Taylor and Gardner, 1963; Minore et al., 1969; Webster, 1978; Heilman, 1981; Lowerts and Stone, 1982). These studies indicate that bulk densities above 1.8 g cm^{-3} generally stop root penetration. Spruce roots seem to be hindered at lower bulk densities than pine roots (Minore et al., 1969; Zisa et al., 1980). At lower bulk densities, but still heavily compacted, roots become short, thicker, crooked, and strongly suberized. But before clearly visible symptoms occur the roots are also hindered, which affects top growth (Mitchell et al., 1981).

In the present study, the aeration and moisture content do not seem likely to have restricted root growth, since moisture content was held within 15%–20% (cf. Eavis, 1972; Warnaars and Eavis, 1972; Stolzy, 1974). Thus, mechanical resistance is probably the critical factor (Wiersum, 1957). Mechanical resistance increases as pore volume decreases (Aubertin and Kardos, 1965; Russel and Goss, 1974), but the resistance is probably more related to the macroporosity than the total pore volume (Van der Weert, 1984; Terry and Campbell, 1981) and the macroporosity will decrease as an exponential function of applied pressure (Eriksson, 1982). In the present study, it is evident that total pore volume does not explain why growth is more impeded in, for example, the coarse sandy till, than in coarse sand (51% and 46% pore volume, respectively, at 400 kPa).

Generally, the impacts of soil compaction seem to be a negative exponential function of soil strength (Taylor et al., 1966). Here the data on root impedence obtained seem to fit quite well to an empirical equation proposed by Russel and Goss (1974):

$$y = a \cdot e^{-bx}$$

where y is the rate of root elongation, when the external pressure is x bar, and a and b are constants. Obtained correlation coefficients show that the proposed equation may also be valid for pine and spruce seedlings (Table 3.5). The impedance factor (b) provides good information on the impact of soil compaction in different soil types.

Based upon these data, a rough estimation of tolerable ground pressure can be calculated. However, shear forces and vibrations may reduce the permissible ground pressure still more (Greacen and Sands, 1980; Eriksson, 1982) and thus also dynamic stresses. Compared with known static ground pressures of logging machines of 50–150 kPa, it is not surprising that deep ruts occur on certain soils (Eriksson, 1981).

Table 3.5 Impedance of root penetration due to soil compaction in five different soil types, based upon the equation $y = a \cdot e^{-bx}$, where y is the rate of root elongation when the external pressure is x bar and b is the impedance factor (Russel and Goss, 1974); based upon obtained growth reduction 16–20 days after sowing.

	Pine			Spruce		
Soil	n	R^2	$-b$	n	R^2	$-b$
Coarse sand	2	0.99	0.042	2	0.80	0.067
Fine sand	1	0.88	0.044	1	0.94	0.075
Coarse sandy till	4	0.94	0.103	4	0.85	0.154
Medium sandy till	2	0.98	0.144	3	0.97	0.250
Fine sandy till	4	0.89	0.152	2	0.96	0.264

Soil compaction introduced by forestry machinery in thinning operations may thus have an impact on tree growth. Even in the transition zone, Wert and Thomas (1981) found a clear retardation of Douglas-fir growth on compacted soil that probably will last throughout the rotation. Surprisingly, Smolander et al. (1981) found that compaction of a till soil did not affect the height growth of pine seedlings. However, they used 3-year-old pine seedlings from the nursery and grew them for only one season into compacted soil. The growth during the first season may be largely determined by the treatment in the nursery the previous year.

The amount of evidence on the impacts of soil compaction justifies more detailed investigations. The soil will recover, but the restoration may take 5–30 years (Hatchell and Ralston, 1971; Holman et al., 1978; Kardell, 1978; Ivanov, 1976, 1980) and as yet we know very little about how the restoration process takes place. However, the soil fauna, vegetation, and ground frost will influence the restoration rate of till soils (Gardner and Telfair, 1954; Thorud and Frissel, 1969; Klimo, 1983). For example, Teepe et al. (2004) measured fluxes of greenhouse gases N_2O and CH_4 across a wheel ruts after the 20 Mg forwarder at three beech (*Fagus sylvatica* L.) forest sites with soils of different texture. Soil compaction in the middle of the wheel rut caused considerable increase of N_20 emissions with values elevated by up to 40+ times the uncompacted ones. Compaction reduced by up to 90% and at the silty clay loam its effect was such was even released. These changes in N_20 and CH_4 fluxes were caused by a reduction in macropore volume and an increase in the water-filled pore space. According to McNabb et al. (2001), a macropore volume of less than 10% is usually considered to be too low to maintain a sufficient supply of oxygen into the soil.

Conclusions

Soil compaction may be one of the reasons for severe growth reduction in trees close to strip roads. More severe growth reduction in spruce than in pine may be caused by higher vulnerability for soil compaction.

References

Aaltonen, V.T., 1923. Über die räumliche Ordnung der Planzen auf dem Felde und im Walde. Acta For. Fenn. 25, 1–85.

Adams, W.A., 1973. The effect of organic matter on the bulk and true densities of some uncultivated podzolic soils. J. Soil Sci. 24, 10–17.

Adams, P.W., Froehlich, H.A., 1981. Compaction of Forest Soils. Oregon State University, Extension Service, PNW-217, 13 pp.

Alexander, E.B., 1980. Bulk densities of California soils in relation to other soil proper- ties. Soil Sci. Soc. Am. J. 44, 689–692.

Aubertin, G.M., Kardos, L.T., 1965. Root growth through porous media under controlled conditions. I. Effect of pore size and rigidity. Proc. Soil Sci. Soc. Am. 29, 290–293.

Bertilsson, G., 1969. Studier över tryckets markpåverkan. [Studies of pressures impact on soil] Lantbrukshögskolan, Inst. för Markvetenskap. Rapp. Jordbearbetningsavd., Uppsala, Nr. 18, 67 pp. (in Swedish).

Bodman, G.B., Constantin, G.K., 1965. Influence of particle size distribution in soil compaction. Hilgardia 36, 567–591.

Bredberg, C.-J., Wästerlund, I., 1983. Wurzel- und Bodenschäden durch Fahrzeuge. Forstwiss. Centralbl. 102, 86–98 (In German, with english summary).

Campbell, R.G., Willis, J.R., May, J.T., 1973. Soil disturbance by logging with rubber-tired skidders. J. Soil Water Conserv. 28, 218–220.

Cruse, R.M., Cassel, D.K., Averette, F.G., 1980. Effect of particle surface roughness on densification of coarse-textured soil. Soil Sci. Soc. Am. J. 44, 692–697.

Davies, D.B., Runge, E.C.A., 1969. Comparison of rooting in soil parent materials using undisturbed soil cores. Agron. J. 61, 518–521.

Dexter, A.~R., Tanner, D.W., 1974. Time dependence of compressibility for remoulded and undisturbed soils. J. Soil Sci. 25, 153–164.

Dickerson, B.P., 1976. Soil compaction after tree-length skidding in northern Mississippi. Soil Sci. Soc. Am. J. 40, 965–966.

Dormling, I., 1975. Photo- and Thermoperiodic Reactions of Scots Pine Seedlings- Can They Provide Criteria for an Early Test? Transfer of Scots Pine (*Pinussylvestris* L.) Seed. Royal College of Forestry, Department of Forest Genetics, Stockholm, Research Notes, 17, pp. 125–140 (in Swedish, with English Abstract).

Dormling, I., 1979. Influence of light intensity and temperature on photoperiodic response of Norway spruce provenance. In: Proc. IUFRO Norway Spruce Meeting, S 2.03. 11-S 2.02.11, Bucharest, pp. 398–408.

Dormling, I., Eriksson, G., Jonsson, A., 1977. Photo- and thermoperiodic reactions of Pinussylvestris and Pinuscontorta. In: Experimentell Genekologi. Swedish University of Agricultural Sciences, Uppsala. Report 27, pp. 48–57 (in Swedish, with English Abstract).

Du Rietz, T., 1968. Huvuddragen av Garpenbergstraktens bergbyggnad (Main Drawings of the Bed Rock in the Garpenberg Area). SGU, Set. C 631, Stockholm, 22 pp. (in Swedish, with English Abstract).

Eavis, B.W., 1972. Soil physical conditions affecting seedling root growth. I. Mechanical impedance, aeration and moisture availability as influenced by bulk drawing levels in a sandy loam soil. Plant Soil 36, 613–622.

Egnér, H., Riehm, H., Domingo, W.R., 1960. Untersuchungen iiber die chemische Bodenanalyse als Grundlage für die Beurteilung des Nährstoffzustandes der Böden. II. Chemische Extraktionsmethoden zur Phosphor- und Kaliumbestimmung. Kungl. Lantbrukshögsk. Ann. 26, 199–215.

Ekström, G., 1927. Klassifikation av svenska åkerjordar (Classification of Swedish agricultural soils). SGU, Ser. C, Nr. 345, Stockholm (in Swedish).

Eriksson, L., 1981. Strip Roads and Damages Caused by Machines when Thinning Stands. Swedish University of Agricultural Sciences, Department of Operational Efficiency, Garpenberg. Report 137, 44 pp. (in Swedish, with English abstract).

Eriksson, J., 1982. Soil Compaction and Root Environment. Swedish University of Agricultural Sciences, Department of Soil Sciences, Uppsala. Report 126,138 pp. (in Swedish, with English Abstract).

Eriksson, J., Håkansson, I., Danfors, B., 1974. Jordpackning – markstruktur- gröda. Jordbrukstekniska Inst., Uppsala Medd 354, 82s. (The effect of soil compaction on soil structure and crop yield. Swedish Institute of Agricultural Engineering. English translation by Kristian Aase. USDA – Agric. Res. Service, 101 pp.).

Froehlich, H.A., 1976. The influence of different thinning systems on damage to soil and trees. In: IUFRO, XVI, Proc. Div. IV, Norway, pp. 333–344.

Froehlich, H.A., 1978. Soil compaction from low ground-pressure, torsion-suspension logging vehicles on three forest soils. For. Res. Lab. Oregon State University, Corvallis. Res. Pap.36, 12 pp.

Gardner, M.R., Telfair, D., 1954. New techniques for the study of restoration of compacted soil. Science 120, 668–669.

Gent Jr., J.A., Ballard, R., Hassan, A.E., 1983. The impact of harvesting and site preparation on the physical properties of Lower Coastal Plain forest soils. Soil Sci. Soc. Am. J. 47, 595–598.

Greacen, E.L., Sands, R., 1980. Compaction of forest soils. A review. Aust. J. Soil Res. 18, 163–189.

Gupta, S.C., Larson, W.E., 1979. A model for predicting packing density of soils using particle-size distribution. Soil Sci. Soc. Am. J. 43, 758–764.

Hägglund, B., Svensson, L.-G., 1982. Site Conditions on Forest Land and Wetlands in Sweden. Swedish University of Agricultural Sciences, Department of Forest Survey, Umeå. Report 35,122 pp. (in Swedish, with English Abstract).

Hassan, A.E., 1978. Effect of mechanization on soils and forest regeneration. I. Coastal plain organic soils. Trans. ASAE 21 (6), 1107–1112.

Hatchell, G.E., 1981. Site preparation and fertilizer increase pine growth on soils compacted in logging. South. J. Appl. For. 5, 79–83.

Hatchell, G.E., Ralston, C.W., 1971. Natural recovery of surface soils disturbed in logging. Tree Planters' Notes 22 (2), 5–9.

Hatchell, G.E., Ralston, C.W., Foil, R.R., 1970. Soil disturbances in logging. J. For. 68, 772–775.

Heilman, P., 1981. Root penetration of Douglas-fir seedlings into compacted soil. For. Sci. 27, 660–666.

Hildebrand, E.E., 1983. Der Einfluss der Bodenverdichtung auf die Bodenfunktionen in forstlichen Standort. Forstwiss. Centralbl. 102, 111–115.

Holman, G.T., Knight, F.B., Struchtemeyer, R.A., 1978. The effects of mechanized harvesting on soil conditions in the spruce-fir region of north-central Maine. In: Bulletin, Life Sciences and Agriculture Experiment Station, University of Maine, No. 751, 13 pp.

Ingelög, T., Olsson, M.T., Bödvarsson, H., 1977. Effects of Long-Term Trampling and Vehicle-Driving on Soil, Vegetation and Certain Soil Animals of an Old Scots Pine Stand. Royal College of Forestry, Department of Forest Ecology and Forest Soils, Stockholm-Uppsala. Research Notes 27, 84 pp. (in Swedish, with English abstract).

Ingestad, T., 1979. Mineral nutrient requirements of Pinus syilvestris and Picea abies seedlings. Physiol. Plantarum 45, 373–380.

Ivanov, B.N., 1980. Effect of the soil moisture of felled areas on the soil density. Lesovedenie (5), 29–36 (in Russian, with English abstract).

Ivanov, B.N., 1976. Dynamics of density of soil in clearings of spruce forests. Lesovedenie (4), 26–30 (in Russian, with English Abstract).

Johnson, M.G., Beschta, R.L., 1980. Logging, infiltration capacity, and surface erodibility in western Oregon. J. For. 78, 334–337.

Johnson, J.A., Hilström, W.A., Miyata, E.S., Shetron, S.G., 1979. Strip selection method of mechanized thinning in northern hardwood pole size stands. In: Michigan Technological University, Ford Forestry Center, Research Note No. 27, 13 pp.

Kardell, L., 1978. Traktorskador och tillväxtförluster hos gran – analys av ett 10-årigt försök (Tractor damages and growth losses in spruce — analyses of a 10 year-old experiment). Sver. Skogsvårdsf'örb. Tidskr. 76 (3), 305—322 (in Swedish).

Klimo, E., 1983. The influence of clearcut logging on soil properties and the cycle of elements in the ecosystem of spruce forest. Lesnictvi 29 (3), 497—512 (English Abstract).

Koolen, A.J., 1974. A method for soil compactability determination. J. Agric. Eng. Res. 19, 271—278.

Köstler, J.N., Brückner, E., Bibelrichter, H., 1968. Die Wurzeln der Waldbäume. Verlag Paul Parey, Hamburg und Berlin, 284 pp.

Laitakari, E., 1927. Männyn juuriristo.Morfologinen tutkimus (The root system of Pine (Pinussylvestris) A morphological investigation). Acta For. Fenn. 33.1, 380 pp.

Laitakati, E., 1935. Koivun juuristo (The root system of birch (Betulaverrucose and odorata)). Acta For. Fenn. 41 (2), 217.

Lowerts, G.A., Stone, J.M., 1982. The effect of soil bulk density on nursery sweet- gum seedling growth. Tree Planters' Notes 33 (3), 3—5.

Lundqvist, G., Hjelmqvist, S., 1941. Beskrivning till kartbladet Hedemora (Description of the Geology Map to Hedemora). SGU, Serie Aa 184, Stockholm 147 pp. (in Swedish).

Lyford, W.H., Troedsson, T., 1973. Fragipan horizons in soils on moraines near Garpenberg, Sweden. Stud. For. Suec. 108, 21 pp.

McCashion, J.D., Rice, R.M., 1983. Erosion on logging roads in north-western California: how much is avoidable? J. For. 81, 23—26.

McIntyre, D.S., 1973. Procuring undisturbed cores for soil physical measurements. In: Loveday, J. (Ed.), Methods for Analysis of Irrigated Soils. Techn. Comm. No. 54. Commonw. Bur. Soils, pp. 154—165.

McNabb, D.H., Startsev, A.D., Nguyen, H., 2001. Soil wetness and traffic level effects on bulk density and air-filled porosity of boreal forest soils. Soil Sci. Soc. Am. J. 65, 1238—1247.

Miles, J.A., Hartsough, B.R., Avlani, P., Chancellor, W.J., 1981. A soil compaction study on the Blodgett experimental forest. In: Forest Regeneration, 81. ASAE Publication 10, Raleigh, NC, pp. 72—81.

Minore, D., Smith, C.E., Woollard, R.F., 1969. Effects of High Soil Density on Seedling Root Growth of Seven NorthwesternTree Species. USDA, Forest Service, Pacific North- west For. Range Exp. Stn., Res, Portland, p. 6. Note PNW-112.

Mitchell, M.L., Hassan, A.E., Davey, C.B., Gregory, J.D., 1981. Effect of soil compaction on root development and seeding establishment. Am. Soc. Agric. Eng. Paper, No.81—1040, 20 pp.

Olsson, M.T., 1985. Micromorphical Analysis of Compacted Soil. Swedish University of Agricultural Sciences, Department of Forest Soils, Uppsala, 10 pp.

Perry, T.O., 1964. Soil compaction and loblolly pine growth. Tree Planters' Notes 67 (7), 9.

Raghavan, G.S.V., Kyes, E., Taylor, F., 1981. Soil compaction effects on soil productivity. In: Forest Regeneration. ASAE Publication 10-81, Raleigh, NC, pp. 90—97.

Rawls, W.J., 1983. Estimating soil bulk density from particle size analysis and organic matter content. Soil Sci. 135, 123—125.

Russel, S.R., Goss, M.J., 1974. Physical aspects of soil fertility.The response of roots to mechanical impedance. Neth. J. Agric. Sci. 22, 305—318.

Sands, R., Bowen, G.D., 1978. Compaction of sandy soils in radiata pine forests. II. Effects of compaction on root configuration and growth of radiata pine seedlings. Aust. For. Res. 8, 163–17.

Sands, R., Greacen, E.L., Gerard, C.J., 1979. Compaction of sandy soils in radiata pine forests. I. A penetrometer study. Aust. J. Soil Res. 17, 101—113.

Scholander, J., 1973. Tear Resistance of Some Forest Soils. Royal College of Forestry, Department of Operational Efficiency, Stockholm. Research Notes 61, 69 pp. (in Swedish, with English Abstract).

Sidle, R.C., Drlica, D.M., 1981. Soil compaction from logging with a low-ground pressure skidder in the Oregon Coast Ranges. Soil Sci. Soc. Am. J. 45, 1219–1223.

Smolander, H., Räsänen, P.K., Kostamo, J., 1981. Effect of soil compaction on transpiration and height increment of planted Scots pine seedlings. Silva Fenn. 15 (3), 256–266 (English Abstract).

Soane, B.D., Dickson, J.W., Campbell, D.J., 1982. Compaction by agricultural vehicles: a review. Ill. Incidence and control of compaction in crop production. Soil Tillage Res. 2, 3–36.

Steinbrenner, E.C., Gessel, S.P., 1955. The effect of tractor logging on physical properties of some forest soils in south-western Washington. Soil Sci. Soc. Am. Proc. 19, 372–376.

Stolzy, L.H., 1974. Soil atmosphere. In: Carson, E.W. (Ed.), The Plant Root and its Environment. University Press of Virginia, Charlottesville, pp. 335–361.

Taylor, H.M., 1974. Root behaviour as affected by soil structure and strength. In: Carson, E.W. (Ed.), The Plant Root and its Environment. University Press of Virginia, Charlottesville, pp. 271–291.

Taylor, H.M., Gardner, H.R., 1963. Penetration of cotton seedling taproots as influenced by bulk density, moisture content, and strength of soil. Soil Sci. 96, 153–156.

Taylor, H.M., Robertson, G.M., Parker Jr., J.J., 1966. Soil strength – root penetration relations for medium- to coarse-textured soil materials. Soil Sci. 102, 18–22.

Teepe, R., Brumme, R., Beese, F., Ludwig, B., 2004. Nitrous Oxide Emission and methane consumption following compaction of forest soils. Soil Sci. Soc. Am. J. 68, 605–611.

Terry, T.A., Campbell, R.G., 1981. Soil management considerations in intensive forest management. In: Forest Regeneration. ASAE Publication 10-81, Raleigh, NC, pp. 98–106.

Thorud, D.B., Frissel Jr., S.S., 1969. Soil rejuvenation following artificial compaction in a Minnesota oak stand. Minn. For. Res. Notes, 208, 4 pp.

Trimble Jr., G.R., Weitzman, S., 1953. Soil erosion on logging roads. Soil Sci. Soc. Am. Proc. 17, 152–154.

Tworkoski, T.J., Burger, J.A., Smith, D.W., 1983. Soil texture and bulk density affect early growth of white oak seedings. Tree Planters' Notes 34, 22–25.

Van der Weert, R., 1984. Influence of mechanical forest clearing on soil conditions and the resulting effects on root growth. Tropical Agric. (BWT) 51, 325–331.

Vater, H., 1927. Die Bewuzelung der Kiefer, Fichte und Buche. Taranter Forstl. Jahrb. 78, 65–85.

Veihmeyer, F.J., Hendrickson, A.H., 1948. Soil density and root penetration. Soil Sci. 65, 487–493.

Warnaars, B.C., Eavis, B.W., 1972. Soil physical conditions affecting seedling root growth.II. Mechanical impedance, aeration and moisture availability as influenced by grain-size distribution and moisture content in silica sands. Plant Soil 36, 623–634.

Wästerlund, I., 1985. Compaction of till soils and growth tests with Norway spruce and Scots pine. For. Ecol. Manag. 11, 171–189.

Webster, D.H., 1978. Soil conditions associated with absence or sparse development of apple roots. Can. J. Plant Sci. 58, 961-~J69.

Weft, S., Thomas, B.R., 1981. Effects of skid roads on diameter, height and volume growth of Douglas-fir. Soil Sci. Soc. Am. J. 45, 629–632.

Wiersum, L.K., 1957. The relationship of the size and structural rigidity of pores to their penetration by roots. Plant Soil 9, 75–85.

Young, R.A., Voorhees, W.B., 1982. Soil erosion and runoff from planting to canopy development as influenced by tractor-wheel traffic. Trans. ASEA 25 (3), 708–712.

Zisa, R.P., Halverson, H.G., Stout, B.B., 1980. Establishment and Early Growth of Conifers on Compact Soils in Urban Areas. USDA, Forest Service, Res. Pap., NE 451.

Further reading

Armson, K.A., Williams, J.R.M., 1960. The root development of red pine (Pinus resinosa Ait.) seedlings in relation to various soil conditions. For. Chron. 36, 14–17.

Fayle, D.C.F., Pierpoint, G., 1975. Evolvement of root and shoot systems of red pine on soils of different moisture availability. In: Bernier, B., Winger, C.H. (Eds.), Forest Soils and Forest Land Management. Les Presses de l'Universit~ Laval, Quebec, pp. 11–25.

Foil, R.R., Ralston, C.W., 1967. The establishment and growth of loblolly pine seedlings on compacted soils. Soil Sci. Soc. Am. Proc. 31, 565–568.

Froehlich, H.A., 1979. Soil compaction from logging equipment: effects on growth of young ponderosa pine. J. Soil Water Conserv. 34, 276–278.

Osen, H.-j., Wästerlund, I., 1989. Terrain and Vehicle Research with Reference to Forestry at the Swedish University of Agricultural Sciences. Department of Operational Efficiency, Garpenberg. Research Notes No 149. (Swedish, Engl. Summary).

Sannikov, Y.G., Barantsev, A.S., 1981. The effect of tractor skidding tracks on the growth of young Scots pine stands. Lesn. Khoz. 1, 14–15 (Ru.).

Wästerlund, I., 1983. Growth reduction of trees near strip roads resulting from soil compaction and damaged roots – a literature survey. Sver. Skogsvårdsförb. Tidskr. 81 (2), 97–109 (in Swedish, with English Abstract).

Youngberg, C.T., 1959. The influence of soil conditions, following tractor logging, on the growth of planted Douglas-fir seedlings. Soil Sci. Soc. Am. Proc. 23, 76–78.

CHAPTER 4

Soil and root strength

Soil strength

Introduction

Increased mechanisation of forest work leads to more intensive traffic in the stand. Shortage of woods from clear-cuttings means that timber from thinning operations is also needed to receive the needed woody biomass for different production uses and not only as a result of tending operation.

Physical facts

In a Nordic forest, the ground floor is usually consisting with a 3–10 cm thick humus layer (Hägglund and Svensson, 1982) that is elastic-brittle. Below that layer there is often a podzol with till soil, which is compacted by a glacial layer that was up to 2 km thick. That heavy load made the ground below 0.5 m depth rather hard to penetrate. Thus, we can concentrate us to the uppermost 10–20 cm where we find the tree roots (Fig. 4.1).

Wheeled or tracked ground contact exerts forces from the vehicle down to the contact with the ground. One force is the normal pressure (see Fig. 4.2) which is mainly the weight per wheel of the machine (e.g., a four-wheeled 6 tons vehicle mean on average 1.5 tonnes is put on the tire load per wheel), which here is one of the lightest harvesters, normally a harvester is 12–24 tonnes. If it is a harvester, the load from the crane will be added on the side where the crane is pointing at. The harvester head may weigh roughly 1 tonne (0.5–1.5 tonne), and the tree in the head may be about 0.5 tonne, thus in summary 3 tonnes will be put on the ground, spread out on the contact area of 70 cm × 25 cm = 1.750 cm^2. Thus, a load of 3000/1750 = 1.7 N per cm^2. The longitudinal stresses may be 1000 N but then the machine must be in total balance, but in single cases it could be 4000 N. Spread out on the contact area it will be 0.29–2.28 N cm^{-2}. Thus, the normal force is normally much higher, and therefore the load of the machine will press the humus layer down and stretch the roots.

Soil and Root Damage in Forestry
ISBN 978-0-12-822070-2
https://doi.org/10.1016/B978-0-12-822070-2.00004-3
© 2020 Elsevier Inc.
All rights reserved.

Figure 4.1 Tree roots in a newly thinned spruce stand and these roots have been passed over with a rubber-wheeled 6 tonnes 4-wheeled harvester. About 10 cm deep in the top floor (= humus layer). Obs: the thicker long lateral has a partial bark removal. *(Photo I. Wästerlund.)*

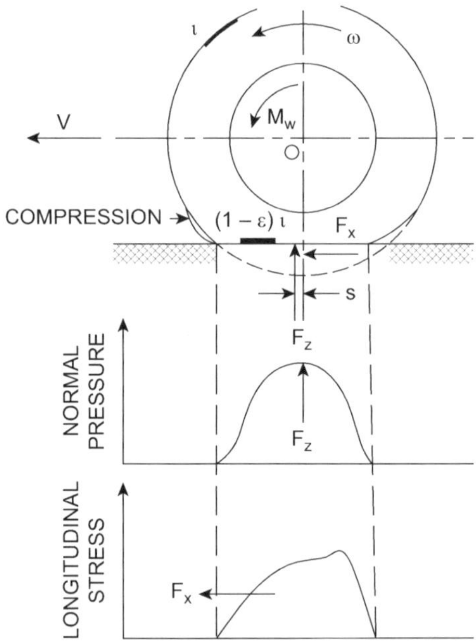

Figure 4.2 Model how the forces are acting beneath a wheel without slippage. *(After Wong, J.Y., 1978. Theory of Ground Vehicles. John Viley & Sons, New York.)*

Nowadays it has become popular to talk about how to avoid rutting after machines working in the forest often without knowing more about what a damage means for the forest but it is fashion (see Chapter 1).

Soil testing is usually done with an annular shear ring that rotates horizontally, or a cone penetrometer is often used in terramechanics to describe soil properties for off-road vehicle performance. A number of equations are being developed to model how a wheel or a track is acting on the ground and acting during vehicle mobility. See further Chapter 6. However, Karafiath and Nowatzski (1978) claim that the shear ring is not confined laterally, and therefore, the failure surface is not necessary the bottom plane of the ring shear plate. Furthermore, roots are twisted around the shear ring which is far from what happens under a wheel!

The design of the machine may be less successful. For example, may a machine with a rigid mechanical transmission (very common on forest machinery, especially on skidders) have a forced wheel slip. The machine may have the same rotational speed on all wheels although different tire inflation pressure and differential on some axles, tire sizes, wear of the tires, the load on the tires, etc., will result in different rolling radius for each one of the tires and thus at least one of them is forced to slip (Olsen and Wästerlund, 1989). The machine transmission design could be worse than that. For example, could the front wheels be made to run faster than the rear ones on purpose to work better steering downhill (Malmberg, 1989). Since very little is known about the strength properties of the forest floor, there is an urgent need for increased research in this topic (Bekker, 1980; Wästerlund, 1989a,b). Preferably the testing tools should be something similar to a wheel, because the tested forces should imitate those acting around a real wheel. We have therefore constructed a new type of test-rig, acting similar to a slipping rigid wheel.

The aim is to analyze the strength properties of the forest floor.

The test-rig

The test-rig acts as a slipping rigid wheel and is powered by two hydraulic cylinders and is carried by a loading machine, which supports the rig with hydraulics and electricity for measuring devices. Load, displacement, tractive force, and sinkage are measured during the test. The tested wheel segments, 30, 60, 120, and 250 mm wide segments, are made of steel with 8–10 mm plain rubber glued to the contact area, and one 60 mm wide segment which is supplied with 20 mm high steel lugs. The radius of the segments is 508–510 mm (Fig. 4.3).

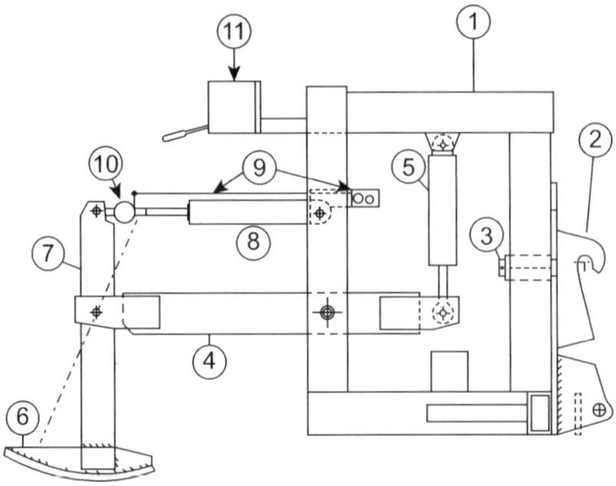

Figure 4.3 Principal drawing of the test-rig for measuring the strength in the forest floor. 1. Frame. 2. Adapter toolholder. 3. Axle spindle. 4. Lever for load application. 5. Hydraulic cylinder for loading. 6. Wheel segment. 7. Wheel spoke. 8. Hydraulic cylinder for loading. 9. Horizontal position indicator. 10. Transducer for horizontal force. 11. Control and settings valves for hydraulic pressures and flow. *(Design by Lars Moberg.)*

Samples from all tested soils have been taken for determination of moisture content (on oven dry basis), organic matter, and for sieving analysis (mineral soil). On forest sites, the thickness of the humus layer and root cross-sectional area are measured as well as the number of roots. Occurrence of stones is noted by depth from surface. In some cases, the humus layer has been removed before a test-run or the humus and the roots have been cut off at the sides of the contact area to see the effect of the absence of those components on the total strength.

Results and discussion

Site F 1

At each site, the sinkage of the wheel segment at only normal load, tractive force during displacement, maximum tractive force, and sinkage during displacement were measured. A number of test-runs with different loads may then give a picture of how the soil behaves under stress. In Fig. 4.4, the sinkage at increased normal load is shown for soil type F 1. Already at a normal load of about 175 kPa, the sinkage is as much as the thickness of the humus layer, and therefore it varies with the humus thickness. On the other hand, there is a remarkable sinkage into the soil without humus layer. Maybe the removal of the roots that was found in the humus layer decreased the total strength of the mineral soil.

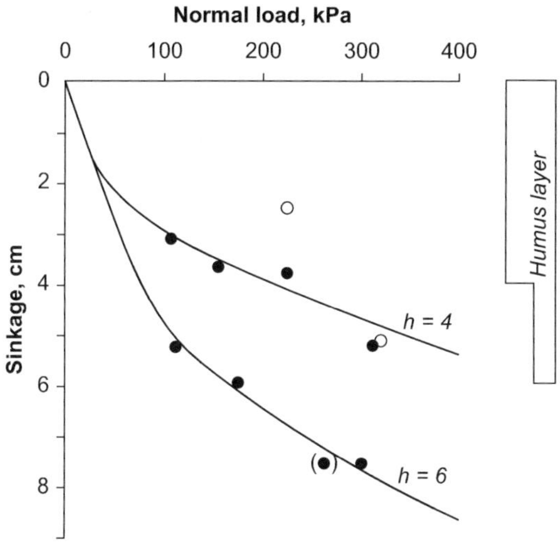

Figure 4.4 Normal load and sinkage relationships at site F1 measured with a 60-mm wide segment with rubber. *h*, thickness of the humus layer; *o*, humus layer removed before test; (*0*), humus layer with roots cut around the foot print area but still on place.

The maximum measured force depends on load up to a certain level (Fig. 4.5). The friction coefficient for the 60 mm wide segment in this case is 0.6. A breakage of the root mat (clearly audible!) resulted here in a reduction of the maximum tractive force compared to what could be expected from the friction line. The maximum extractable force appears to be around 16 N cm^{-2}.

The sinkage during displacement is very little at low normal loads but increases rapidly at high loads (Fig. 4.4). A low root cross-sectional area seems to imply a high sinkage. The cross marked F in figure represent a place with a root-sectional area of only 0.12 cm^{-2}. The average for other tested places at site F1 was 0.40 cm^{-2}. In summary, the results of the tests from site F 1 indicate a soft floor with low strength. A subjective inspection of the site should tell the experienced soil scientist that a heavy forwarder (20—30 tonnes) may create big ruts in this stand.

In summary, the results of tests from site F 1 indicate a soft forest floor with low strength. A subjective inspection of the site should tell the experienced soil scientist that a heavy forwarder may create big ruts in this stand.

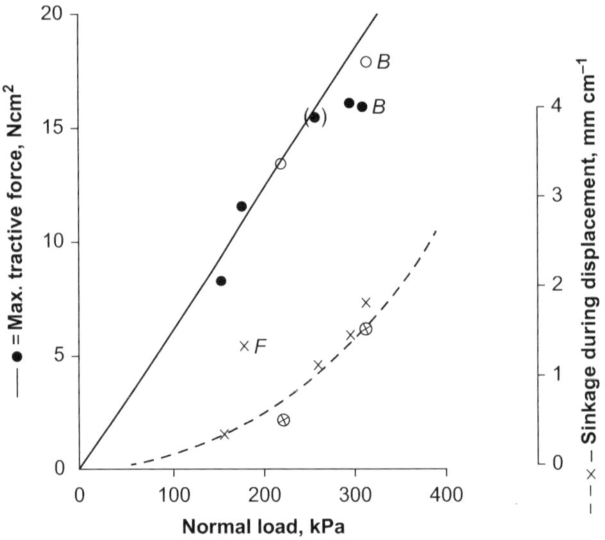

Figure 4.5 Maximal tractive force and sinkage during displacement at different normal loads *B*, breakage; *F*, few roots. For other explanations see Fig. 4.4.

Comparisons between all sites

1. Load and sinkage

Load and sinkage relationships for all tested sites are presented in Figs. 4.6 and 4.7. The sand sediments have a high load carrying capacity, whereas it is low for the silty sand (C) maybe because of its high organic content and low degree of adhesion. Sinkage at the forest site seems to depend mostly upon the thickness of the humus layer. However, it should be pointed out that the sinkage is measured under load. When lifting up the wheel segment, some recovery will occur of rut formation because of the elasticity of the humus layer. Some of the curves have been obtained using different wheel segments (e.g., at A, E, and F). In all these cases, the normal load seems to explain the major part of the sinkage curve and the width of the wheel segment plays no other role than changing the foot print area. However, this was not found to be true for the peat bog (D). Here, the different widths formed their own sinkage curves. It seems as if the total load plays the major role and it is only partly modified by the width of the contact area ($r^2 = 0.92$ according to the multiple regression analysis). Maybe the reason for this behavior depends on the low bearing capacity of this material. These results for peatland deviate somewhat from the results from muskeg obtained by Wong et al. (1979).

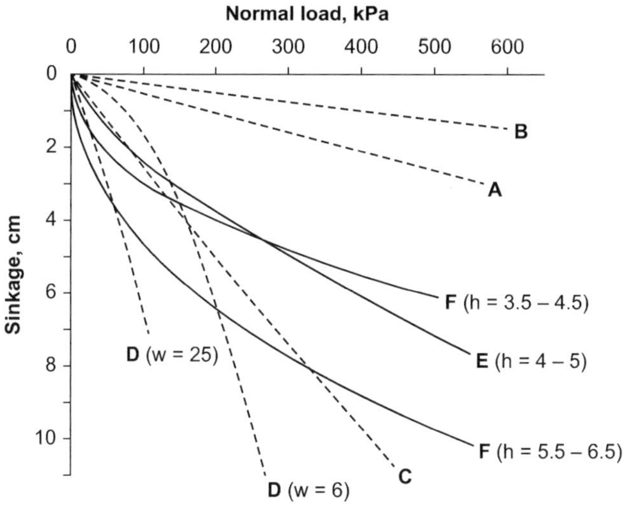

Figure 4.6 Load and sinkage relationships in different soils (A–E, described in Table 4.1). *h*, thickness of the humus layer in cm; *w*, width of the wheel segment in cm.

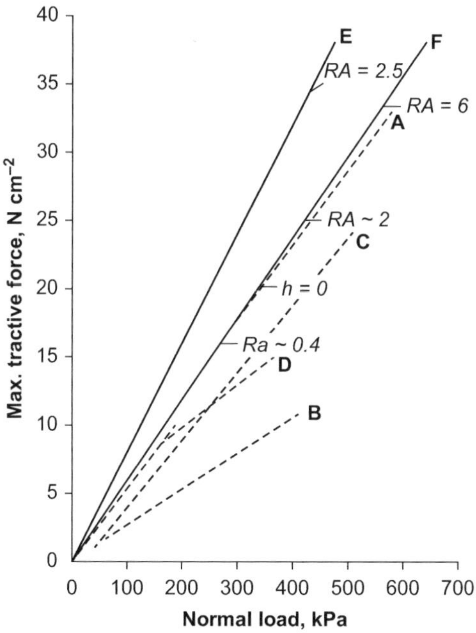

Figure 4.7 Maximum tractive force with different normal loads for the tested soils. *RA*, cross-section area of the sum of roots in the foot print.

2. Tractive force

The maximum tractive force varies also with soil type. In general, it appears as if the tractive force increases almost linearly with increased load regardless of width of rubber supplied wheel segments. However, the results from the peat bog (D) do not follow this general trend. At normal loads above a 150 kPa the traction appears to be less efficient indicating that traction forces above 9 N cm^{-2} should be avoided in order not to destroy that mat completely. The traction data for the peat bog appear to be similar to those for muskeg during rubber-mat shearing (Wong et al., 1979).

The wheel segment with grousers had usually a better grip than the rubber supplied segments. The friction coefficient could be 0.7–0.9 compared with 0.5–0.65 for the ones with rubber. The regression lines shown in Fig. 4.8 are based on the segments with rubber. The regression line for beach sand (B) indicates a very low friction coefficient for that material. This must be questioned since practical experiences with torque measurements on a forwarder towing another forwarder (Friman and Eriksson, 1989) on the same beach indicate a friction line more similar to sand type A.

At the forest sites, the maximum tractive forces could follow the regression line up to very high values for load and force. However, after a

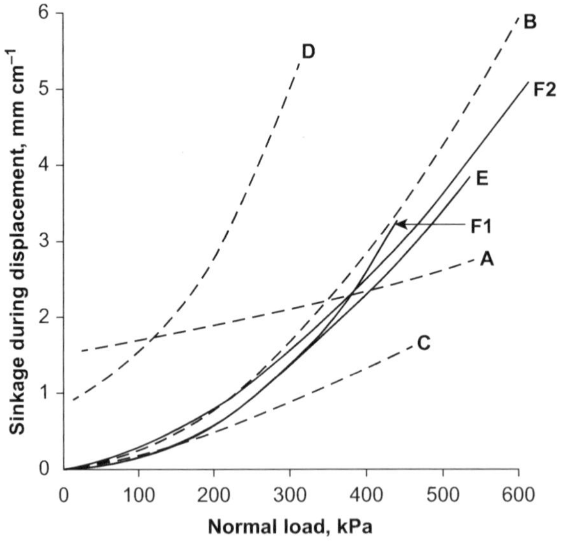

Figure 4.8 Sinkage during displacement of the loaded wheel segment.

breakage of the root mat, the wheel segment sank deep into the soil and the tractive force decreased. This is indicated by the hooks on the E and F lines in Fig. 4.8. On a soil with high bearing capacity as at the E site, root mat does not seem to influence so much on the total strength. At a softer soil as at the F sites, different amounts of roots appear to strongly influence the strength of the forest floor. The net of roots in the foot print area varied from plot to plot and the amount could be from 150 to 1000 tree and shrub roots or rhizomes per square meter. Lowest amount of roots was found on site F 1. Highest cross-sectional area was 13 cm^2 recorded at one F2 plot.

According to a literature review in Chapter 2, the tensile strength of roots could be 1–5 kN cross-sectional area. Root extraction force measurements by Anderson et al. (1989) gave an average of 0.5 kN cm^{-2} for ca Sitka spruce roots. For the interpretations of the root mat on the strength of the forest floor value of 15 N cm^{-2} (see Table 4.2 in next section) can be used in the calculations. The cross-sectional area of 0.4, 2, and 6 cm^2 will then contribute to a strength of 2, 10, and 30 N cm^{-2} for the average foot print area 1750 cm^2. The calculations indicate that the root mat may have a major influence on the shear strength of the forest floor.

The maximum tractive forces are obtained at different displacements of the wheel segments. At the sand sites (A and B), the tractive force curve usually raises fast in the beginning of the displacement and then continues to raise a little with increased displacement as can be seen in ordinary text books (cf. Wong, 1978; Yong et al., 1984). At the forest sites, the curve could look similar to the sand curves at low normal loads (<200 kPa). However, at high loads the maximum tractive force was usually obtained within few centimetres of displacement and after that decreased. In these cases, the wheel segment sank rapidly into the soil. Finally, the curves from the silty sand (C) and the peat bog (D) always showed a peak for the tractive force within a few (<4–5) centimetres of displacement indicating a brittle material.

Sinkage during displacement

It appears to be two types of sinkage during displacement. Sand (A) and peat (D) are pushed away or give away at any load when the wheel segment moves. This is shown also with other types of soil testing equipment (cf. Yong et al., 1984). The beach sand should behave in this way too but no

Table 4.1 Brief description of tested sites and soils.

Notation	Site	Soil types	Organic matter %	Moisture content	Number of compl. test-runs
A	Glaciofluvial sand without vegetation	Fine-medium sand, s 80% 0.06–0.6 mm 20% > 0.06 mm	2.0	22–27	20
B	Lake beach without vegetation	Sand, s 6% 0.6–2.0 mm 88% 0.06–0.6 mm	–	17	5
C	Potato land without vegetation	Silty sand, s 75% 0.06–0.6 mm 21% 0.002–0.06 mm	5.9	32	7
D	Cultivated peatland Straw stubble	Well-humified peat >1 m deep	85	220	14
E	Spruce forest 100 yr old Blueberry and moss vegetation	Podzol, sand overlaying, m 10% 0.6–2.0 mm 67% 0.06–0.6 mm	7.4	28–47	20
F1	Thinned 50 yr old spruce forest. Grass and moss vegetation	Podzol. Fine sandy, m. 15% 2–20 mm 51% 0.06–2 mm 34% > 0.06 mm	3.9	20	7
F2	Thinned 45 yr old spruce forest. Grass and moss vegetation	Podzol, fine sandy, m. 12% 2–20 mm 41% 0.06–2 mm 41% 0.02–0.06 mm 6% > 0.002 mm	9	20–35	19

Moisture content is given on oven dry basis. Soil types *s*, sediment; *m*, moraine.

Table 4.2 Compilation of root tensile strength data.

	Diameter, mm under bark	Strength N/mm^2	Extension %	No. of samples
Pine, Pinus sylvestris				
In the stand	1.9–6.0	16.09	8.37	20
12 years old	1.8–3.5	17.17	16.94	9
26 years old	2.0–2.9	12.32	8.7	10
143 years old	2.0–6.8	27.65	8.79	8
W. avege				
Spruce, Picea abies				
Forest soil	1.4–5.1	38.92	27.87	12
Agricultural soil	1.4–5.4	25.10	17.86	20
Forest stand			20.14	20
W. avege		30.245		
Birch				
Betula pendula	1.2–4.6	23.81		
Betula pubescens	1.6–4.8	25.81	8.93	18
W. avege		24.78	10.10	17
Shrubs, rhizomes				
Calluna vulgaris	2.5–6.8			10
Ledum palustre		28.04	8.51	10
Vaccinium myrtillus	2.1–6.5		5.35	10
Vaccinium uliginosum		24.96	6.11	12
W.avege			5.17	

W. avege, weigthed average, based on number of samples.

explanation for this behavior is found here. The other type constitutes of the forest sites and the brittle silty sand (C). Here a slippage at low normal loads causes almost invisible further sinkage.

The other type constitutes of the forest sites and the brittle silty clay sand (C). Here a slippage at low normal loads causes almost invisible further sinkage. For example, one experiment at site E with 8 times repeated loading (175 kPa) and slipping (4 cm per time) at the same spot caused only totally 1.1 cm further sinkage from the first run. On the other hand, the sinkage during displacement may be considerable at high loads. An interesting observation was that these two types of sinkage behavior occurred more or less with the grouser supplied wheel segment. The load or tractive force where the sinkage during displacement becomes apparent may be the threshold value which the forest machinery should be designed for as a maximum value in order to reduce damages in the stand! It may be worthwhile to try to determine such a threshold value during further studies.

Concluding remarks

The used methods, the test-rig, and the results appear not to be fundamentally wrong. Although quite work intensive methods, it seems worthwhile to continue with the measurements. They could also be supplemented with soil density measurements for analysis of the effect of load and slippage on the mineral soil under the humus layer which now cannot be controlled.

The results from these measurements indicate also an explanation to a sudden bog down of a forest machine: The root broke on a soft soil! When the root mat broke, less tractive force could be taken out in the ground contact and at the same time the wheels will sink down into the soil during slippage. This in turns means higher rolling resistance and continued slipping.

It appears also to be wise to have slip control on heavy forestry machines. The forest floor is quite soft giving some rolling resistance but strong enough to allow quite many passages if not the root mat is teared off. Then the rut formation increases rapidly (Fig. 4.9).

Figure 4.9 Deep rutting in the wet humified soil and lots of roots cut off after a thinning operation. *(Photo I. Wästerlund.)*

Root strength

Stumps and roots have similar types of wood cells as the stem but they are thinner than in the stem (Hakkila, 1989). When modeling strength properties of wood components, it is common to just calculate the cross-sectional area. According to Hooks law, the normal stress is

$$F/A = kl/A\sigma = kl/A\gamma$$

which is also called Young's modulus, where k value (spring constant) must be determined experimentally for each material and is the length of the sample (l). The force (F) is linear regardless of form (A), the main thing is to have control of cross-sectional area. The force is assumed to be distributed uniformly over the cross-sectional area (A) and is oriented normal to the cross-sectional area. The shape of the cross-sectional area may be circular, elliptical, or any other geometry, so long as it remains constant along the length of the rod (Bodig and Jayne, 1982). Thus, for roots the main factor is actually to have control of the cross-sectional areas, because they do not change shape during the tensile pull test.

The normal pressure behind a wheel is most often higher compared to the longitudinal stresses in the soft-brittle humus layer, which means the roots are subjected to stretching force since the roots are attached to the soil further away with their branching. This was obvious when I pressed down my loading foot and birch roots 1 m away pulled up the humus layer and during my experiment was pulled out from the soil. The problem was to convince forestry people about it since they were mainly looking up to the crown.

Facts

Systematic studies were started in the eighteenth century. With a simple excavation technique Hales (1727) dug out root systems of cultivated crops and determined their morphology, as well as weight and length. Du Hamel Du Monceasu (1764/65) ex-cavated tree root systems. Then for over hundred years no important root research was documented (Böhm, 1979). However, he missed the Swedish studies by Björkhem, Lundeberg and Scholander (1975) had done both root excavation and soil strength testing and they used the monolith method. Some roots have been studied after final cut of trees (O'Loughlin, 1974; Burroughs and Thomas,1977) or in slopes (Endo and Tsuruta, 1969), and these studies showed rather weak roots.

Root properties

This means that tensile strength and elongation are very important including the strength of bark. The strength of the roots was measured with a Hounsfield tensile tester with the ends of the root samples wired with soft sandpaper to get a good grip. The elongation during each tensile test was measured with a clamp near the grip holder and connected to an extensometer to give the elongation in mm and then calculated in % based on the free length of the sample between the holders. The strength was calculated based on the cross section of the area at where the root broke (Fig. 4.10).

Strength of tree bark

The strength of bark has been studied by Wästerlund (1986). Figs. 4.11 and 4.12 shows the principal design of the test-rig to measure the strength of bark on the roots. By cutting in different ways the properties of bark could be measured (Fig. 4.13) like tangential or longitudinal way, or with total surrounded cutting the adhesion to wood sample could be measured.

Figure 4.10 Hounsfield 5000 tensile tester and extensometer to the right in the figure and on top the 5000 N load cell. *(Photo I. Wästerlund.)*

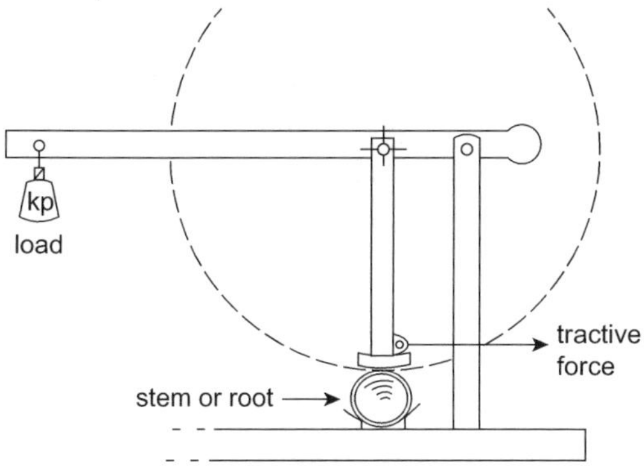

Figure 4.11 Principle drawing of test equipment for measuring the strength of bark.

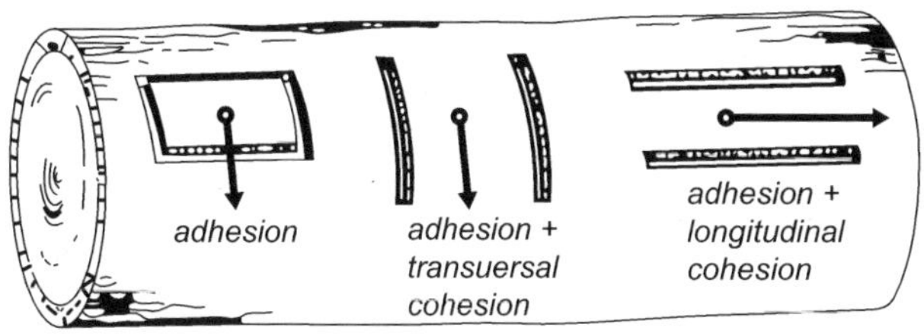

Figure 4.12 By cutting the bark in this way, adhesion to the wood or cohesion could be tested in the test equipment, and by sampling at different times its strength could be tested during different months. *(Drawing S. Falck.)*

Suppose a driven tire is passing a root. The root could be a spruce heart root lying almost uncovered on top of the soil. The wheel is entering the root and is resting only on the root with the lugs. The entry angle is 45 degrees. The contact area between the lugs and the root is 40 cm^2 (cf. Fig. 4.14). Suppose only the load of the tire is influencing the root. The resultant of the force exposure affecting the root is equal to the loads Q of the tire. The shear force F acting on the root bark is then $F = Q \cos 45$.

This may happen in September. To remove the bark, the shear forces must be more than 60 N cm^{-2}. The load should therefore be heavier than $60 \times 40/0.71 = 3,380$ N. The friction coefficient between the tire and the bark could be 0.6. This means that total loads above 5630 N will cause removal of the root bark.

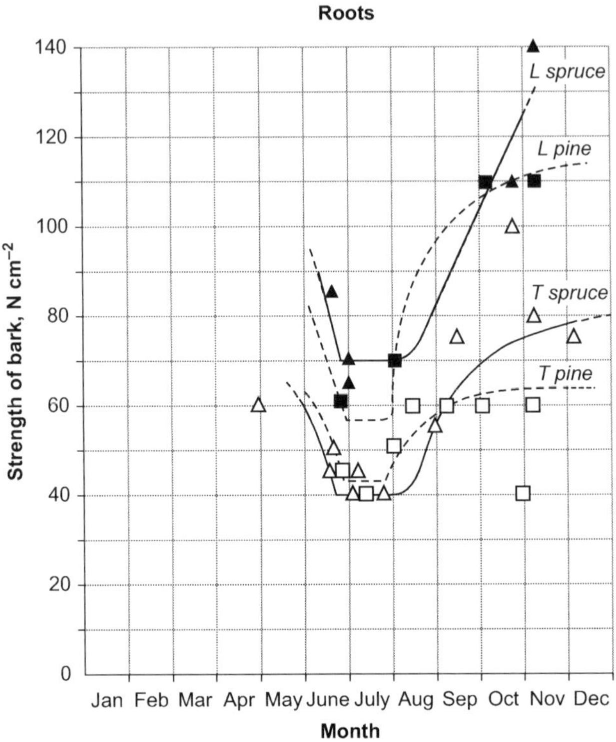

Figure 4.13 The seasonal variation of the strength of bark for roots (N per cm^{-2} Spruce roots (Δ) tangential and (▲) longitudinal. Pine roots (□) Tangential and (■) Longitudinal direction.

Suppose the tire has the dimension 500 × 22.5". With a supposed sinkage into the soil of 15%, this tire may have a total ground contact area of $50 \times 56 = 2800\ cm^2$. That tire should carry a load of less than 5630 N. The average ground pressure will be less than $5.63 \times 10/2,800 = 20$ kPa.

Apparently, such low average ground pressures are hard to achieve with forestry machines and common types of tires. However, the example shows the importance of the load being distributed on a large contact area at the passage of single roots.

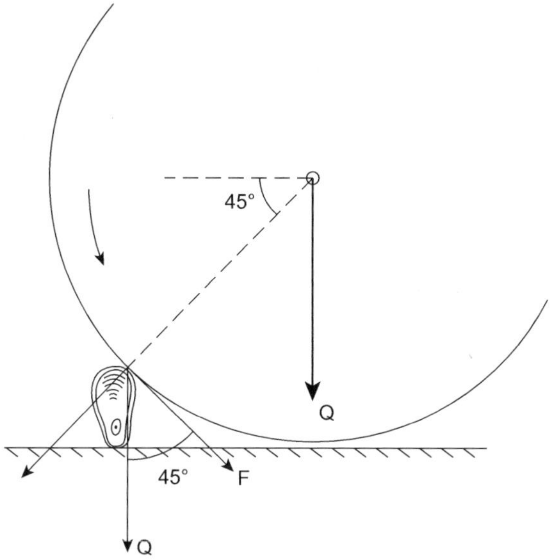

Figure 4.14 An example of the loading situation when a wheel is rolling over a tree root. *(Drawing Wästerlund, I., 1986. The Strength of Bark on Scots Pine and Norway Spruce Trees. The Swedish University of Agricultural Sciences, Department of Operational Efficiency, Garpenberg. Report No 167.)*

References

Anderson, C.J., Coutts, M.P., Ritchie, R.M., Cambell, D.J., 1989. Root extraction force measurements for Sitka spruce. Forestry 62 (2), 127−137.

Bekker, M.G., 1980. Evaluation of Soil/vehicle Relationtionships to Lessen Damage to Forest Road and Off-Road Surfaces.A Literature and State-Of-The-Art Survey.Forest Survice. USDA. Equipment Development Centre, San Dimas, California.

Björkhem, U., Lundeberg, G., Scholander, J., 1975. Root distribution and compressive strength in forest soils. Royal college of forestry, Dept. Forest ecology and forest soils, Stockholm. Research notes no. 22.

Böhm, W., 1979. Methods of studying root systems. Springer-Verlag, Berlin, Heidelberg, New York.

Bodig, J., Jayne, B.A., 1982. Mechanics of wood and wood composites. Van Nostrand Reinhold Company, New York, Cincinannati, Toronto, London, Melborne.

Burroughs Jr., E.R., Thomas, E.R., 1977. Declining root strength in Douglas-fir after felling as a factor in slope stability. In: USDA Forest Service, Intermt. For. And Range Exp. Stn. Utah Res Pap INT-190, Ogden.

Du Hamel Du Monceasu, H., 1764/65. Naturgeschichte der Bäume. Teil I und II. Winterschmidt, Nürnberg.

Endo, T., Tsuruta, T., 1968. On the Effect of Tree's Roots upon the Shearing Strength of Soil. Annual Report of the Hokkaido Bramch Forest Exp. Stn., Sapporo, pp. 167−182 (Engl summary).

Friman, E., Eriksson, U., 1989. Tractive Forces and Torque Distribution on a Forwarder Equipped with an All Hydrostatic Transmission. The Swedish University of Agricultural Sciences, Department of Operational Efficiency, Garpenberg. Uppsatser och resultat nr 154. (Swedish, Engl. Summary).

Hägglund, B., Svensson, L.-G., 1982. Site conditions on forest land and wetlands in Sweden. Swedish University of Agricultural Sciences, Dept. of Forest Survey, Umå. Report 35, p. 122 (in Swedish, with English Abstract).

Hakkila, P., 1989. Utilization of residual forest biomass. Springer-Verlag; Berlin, Heidelberg, New York, London, Paris, Tokyo, Hong Kong.

Hales, S., 1727. Vegetable statisticks. London.

Karafiath, L.L., Nowatzski, E.A., 1978. Soil mechanics for off-road vehicler engineering. Trans. Tech Publications, Claushal, Germany.

Malmberg, C.-E., 1989. The off-road vehicle. In: The Joint Textbook Committee of the Paper Industry, vol. 1. TAPPI, CPPA, Atlanta.

O`Loughlin, C.L., 1974. A study of tree root strength detoriation following clear-felling. Can. J. For. Res. 4, 107−113.

Olsen, H.J., Wästerlund, I., 1989. Rerrain and vehicle research with reference to forestry at Swedish University of agricultural sciences. Dept. of Operational Efficiency, Garpenberg. Research notes no. 149. (Swedish, Engl. Summary).

Wästerlund, I., 1986. The Strength of Bark on Scots Pine and Norway Spruce Trees. The Swedish University of Agricultural Sciences, Department of Operational Efficiency, Garpenberg. Report No 167.

Wästerlund, I., 1989a. Are Present Forestry Machines Smooth Terrain-Machines? SFM Meddelande, pp. 21−34. Stockholm, Stockholm. Nr38.

Wästerlund, I., 1989b. Strengths components in the forest floor restricting maximum tolerable machine forces. J. Terramechanics 26 (2), 177−182.

Wong, J.Y., 1978. Theory of Ground Vehicles. John Viley & Sons, New York.

Wong, J.Y., Garber, M., Radforth, J.R., Dowell, J.T., 1979. Characterisation of the mechanical properties of muskeg with special reference to vehicle mobility. J. Terramechanics 16 (4), 163−180.

Yong, N., Fattah, E.A., Skiadas, N., 1984. Vehicle Traction Mechanics. Elsevier, Amsterdam.

CHAPTER 5

Rutting on peaty soil

Introduction

Swedish forestry area has been used in many ways, for example, to produce trees/timber for forest industry or house building, but also as landscape for recreation, berry picking, mushroom picking, hunters and fishermen when reaching their waters, and Laplanders for their reindeer management. Many of these activities have their own wishes on how the forest should look like, but all agree upon that soil rutting damage after harvest is a bad thing. The need to reduce soil damage is great, especially on frequent traficated areas, and at the same time, forestry vehicles tend to increase in size (more than 20 tonnes unloaded) and increase (Lull, 1959), all with the ambition to increase the productivity and make the operation cheaper (Fig. 5.1).

Changes in soil depth (increased density) have been noticed mainly in agriculture being caused by axle load (Håkansson, 1990). Håkansson (1990) formulated his theories on more homogenous soils than commonly occurring in forest soils compared with agricultural soils. Studies on forest soils have shown similar tendency (Table 5.1). Compare that we still after 10,000 years since the inland ice still can measure the compaction from the 100 m thick ice layer at about 100 cm depth in the ground. On grounding a well-bedded path with logging residues (branches), the soil damages were halved with 20 cm thick layer of branches (Eliasson and Wästerlund, 2007).

Normally, when cutting occurs on weak ground, logging residues are used to protect the ground, but if the logging residues are used for bio-energy, this protection will be missing (Eliasson and Wästerlund, 2007). Periods with frozen ground is possible to use, but they tend to be shorter with the ongoing climate change up in the north.

Root damages at thinning operations may cause both reduced growth and increased wind throwing in addition to increased risk for root rots (Froehlich, 1989; Wästerlund, 1992). In forest soils, there is a risk for permanent damages in the ground, especially on moist soil with reduced bearing capacity, which are difficult to repair (due to roots and stones).

Soil and Root Damage in Forestry
ISBN 978-0-12-822070-2
https://doi.org/10.1016/B978-0-12-822070-2.00005-5
© 2020 Elsevier Inc.
All rights reserved.

Figure 5.1 The tested area with lotted tracks now testing a Ponsse forwarder after one passage.

Table 5.1 Compilation of some older SLU soil studies in Sweden on forest soils and measured soil changes at depth in cm.

Vehicle	Total mass, tonnes	Axle load, tonnes	Rut depth, cm	Density changes maximum depth, cm	Study
Cleaning machine	5–7.5	2.5 –3.5	1–3	ns	Jansson and Wästerlund (1999)
Cleaning machine	11	5.5	2.4–4	ns	Wästerlund (1988)
Forwarder	20	11.4	15	36	Jansson and Johansson (1998)
Excavator	20	Tracks	5	35	Jansson and Johansson (1998)
Forwarder	35	23	9.3	>40	Eliasson and Wästerlund (2007)

ns = not studied.

Those damages could be of three types: soil compaction, rutting, and root damages (Froehlich, 1989). Those damages could be separate or combined (Wästerlund, 1994) and occur both on trees and seedlings. Soil damage and rutting is common after intensive traffic with heavy vehicles on soil with low bearing capacity. The risk for rutting increases with increased traffic (Björkhem et al., 1983; Jansson and Johansson, 1998). Soil compaction occurs usually during the first passages (Hadas, 1994; Carter et al., 1999). In North America, the opinion to soil damages is due to traffic intensity (Lull, 1959; Carter et al., 1999; Klepac et al., 1999). Another drawback with rutting is that the fuel consumption increases with increased depth of rutting (Bygdén et al., 2003).

In Sweden, soil compaction is mostly studied after thinning and cleaning operation (Jansson and Wästerlund, 1999), but not much is done after clear fellings. In North America, soil compaction is seen as a problem (Klepac et al., 1999). Wågberg (2001) studied soil damage after clear felling operations, and it is noted that traffic could locally cause very deep and serious rutting, and the deepest rutting occurred on places where no logging residues were on the ground. A 20 cm deep layer of slash (especially thick branches) considerably reduced soil damage after five passages (Eliasson and Wästerlund, 2007).

Soil damage after a clear cut may to some extent be ameliorated after site preparation according to a study by Gyldberg (1993), who found that driven disc trenching (powered rotation of discs) loosen up soil down to 20 cm below the teeth of a disc trencher. However, this amelioration depends on what site preparation method is used (Ritari and Lähde, 1978), and still it is not known to what extent site preparation may ameliorate compacted soil. In the same way with logging residues actively reinforce ground. Also some bridge constructions or strong geonets (>40 kN m^{-1}) could be used to protect the ground (Bjarnert, 2010).

Demands on high productivity for forwarders cause manufacturers to make bigger ones or they became overloaded, since increased load per turn increases productivity (Brunberg, 2004), which will increase the load on the rear axle. However, an ATV could cause soil rutting due to four-wheel drive without differentials and at uneven loading have forced slippage!

Internationally, trafficability (Anon, 1968) is defined as the ability to pass an area a certain amount of times without soil damages occurring. Soil compaction may damage a number of aspects in the ground like drainage, infiltration of rain, erosion, and mercury leakage from the area and eutrophication of nearby lakes or streams (Greacen and Sands, 1980), and the effects could last for a long time (Alakukku, 2003). Root damages and soil compaction may cause a 50% growth reduction of the edge trees after a bad operation (Wästerlund, 1994).

An analysis of some older data from Swedish University of Agricultural Sciences (SLU) on forestry soils indicates that a vehicle with calculated ground pressure below 50 KPa (according to the old RTG formula, see later on) and an axle load of less than 10 tonnes should be acceptable from soil damage point of view (Wästerlund, 1988) (Table 5.1). The question is how to solve that problem from effective transport of timber point of view in forest terrain.

ElForest is an interesting concept as the load is spread out on two axles. Furthermore, the wheels are following each other in the curve, which means that side slippage of bogies is avoided and thus deep ruts are avoided when curve driving (Edlund et al., 2012). Its drawback was that the whole load was resting on only one axle in uneven terrain (no suspension on the middle axle), which caused the closedown of that project.

Aim

The aim is to compare forwarders with different types of ground contact devices concerning rut formation and soil changes in the depths when driving straight forward. Forwarders were chosen because they are the most heavy machines when fully loaded (up to > 40 tonnes loaded) and are usually the machine type that makes most soil damages in the forest.

Material and methods

The experiments were conducted during four in days in September 2008 to 2011. The rutting was conducted on an agricultural and cultivated peat land, thus an organic soil with decomposed agricultural residues situated 7 km southeast of Tierp (N 60.311 464; E 17.612 218).

The wheel ruts were measured on a straight lane. The vehicles were propelled straight forward with a speed of 3–4 km forward 10 times on a 30 m long test track with measurement of wheel ruts every second meter. The tests were done both without load and with 75% of maximum loads on the machine. At each passage, the wheel rut depths were measured. The weight of the machines was controlled with scales under each wheel with a 75 kg driver and full fuel tank.

The following machines were used:

Lightlogg with tracks and three articulated machine parts weighing empty 13,000 kg. It was loaded with 9750 kg spread out on two vehicle parts. Total weight was 34,370 kg.

ProSilva, a tracked machine with a mass of 23,100 kg. It was loaded with 11,250 kg, and then total mass was 34,370 kg.

Rottne F13S had a mass of 19,520 kg and loaded with 9750 kg and then the total mass was 29,270 kg. Also it runs like Harrier, thus with the rear part running on the side of the front wheels.

Valmet 860.3 when loaded had a weight mass of 30,210 kg. It was loaded with 10,500 kg timber, and it was the test vehicle and control between the years.

Valmet 860.3 with EcoTracks weighs 22,390 kg when unloaded. The tracks are for traction.

Valmet 860.3 with EvoTracks being a mixture between traction and for support. This time, the vehicle when unloaded had the weight of 22,590 kg including the tracks.

Valmet 860.3 had the mass of 22,710 kg with EcoMagnum bogie tracks. These tracks are for support. Wheeled versions were also tested with three different levels of inflation pressure: 6, 3.5, and 2.7 bar. All pressure levels were accepted by the tire manufacture.

Ponsse Visent with eight wheels had empty mass of 16,520 kg including tracks and an extra pair of wheels hanging in the air.

In Ponsse Visent with 10 wheels with EcoMagnum tracks on all wheels, the width of the tracks were 916 mm in the front and 1023 mm on the rear. When empty, the vehicle had the mass of 20,570 kg, including tracks and the fifth axle running. It was loaded with 9000 kg.

Finally, ElForest with four electric motors in four wheel hub, six wheels, and major suspension front and rear and four-wheel steering had 10 tonnes load at 75% level with 15.7 tonnes of its own weight.

Reason for loading to 75% of maximum capacity was to be comparable with older tests done earlier on the same type of peat soil. At older tests, the machines then had much more narrow tires and thus sank deeper down.

The study was made as a comparative study with lotted tracks and wheel ruts were measured after 0, 1, 3, 5, and 10 passages. All wheels on the ground were powered. Measurement of rut depth was made up to a rod placed across the rut and down to deepest position at marked places in advance at 2 m distance. All together 10 measurements were done after putting together left and right side per passage. Then the average of these two sides measurements are given as average per passage.

The studies were done on cultivated organic soil were barley (D in Chapter 4), wheat, and oat on that the soil had been grown and had been used at least for 20 years (D in Chapter 4). The soil consisted of an organic layer circa 20 cm thick overlaying 10 cm layer of fine sand (formerly a lake), and below was a fine clay (depth 25–30 cm). Soil moisture content was between 54% and 67% ± 4% (MC dry) between the years and at different parts of the area, thus surprisingly with little variation.

Measurements of cone resistance was done with Eijkelkamp logger using a standard ASAE cone, thus 30° cone with the base 2.1 cm and basal area 3.2 cm^2 and penetration rate 3 cm s^{-1} (Anon, 1968). Cone resistance was measured at every second point where rut depth measurement was done and was stopped when resistance became more than 1.5 MPa; resistance was recorded at every cm of depth. Two controls were made across the whole area, revealing a homogenous soil (cf. Fig. 5.14).

Pressure in the ground was measured 2011 with load cells (Arvidsson et al., 2001). The vertical pressure in the ground was measured by placing sensors with a liquid pressing on pressure sensors (Figs. 5.2 and 5.3). These sensors were placed at certain depth from a ditch on the side of the track.

Five of the vehicle constellations passed this ditch except the inflation pressure 2.7 bar since the sides of the ditch broke (the last test).

Soil movement is to some extent elastic in organic soils, and the other part is plastic (cf. Fig. 5.11). However, in organic soils, the plastic type of soil deformation is quite small and therefore the test soil was less good in that respect.

At rut depths >15 cm, the driving tests were stopped to avoid destruction of the agricultural soil.

In Table 5.2, the compilation of the vehicles and the ground contact devices is given for the vehicles that were tested.

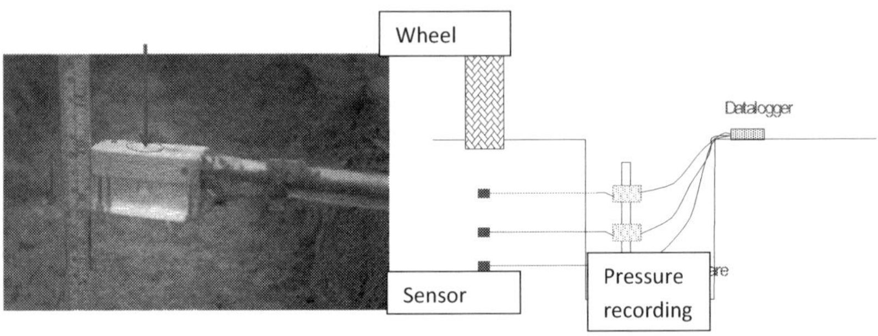

Figure 5.2 The pressure sensor, with the arrow point on the load cell, placed 100 mm down into the soil and 300 + 500 mm down. *(After Arvidsson, J., Trautner, A., van den Akker, J.J.H., Schønning, P., 2001. Subsoil compaction caused by heavy sugarbeet harvesters in southern Sweden II: soil displacement during wheeling and model computations of compaction. Soil Till. Res. 20 (1–2), 79–89.)*

(A) (B)

Figure 5.3 (A) Measurement of pressure and movement in the field. (B) Reinforcement of the sides of ditch was necessary when the loaded forwarders passed due to soft soil. *(After Arvidsson, J., Trautner, A., van den Akker, J.J.H., Schønning, P., 2001. Subsoil compaction caused by heavy sugarbeet harvesters in southern Sweden II: soil displacement during wheeling and model computations of compaction. Soil Till. Res. 20 (1–2), 79–89 + Photo Wästerlund.)*

Ground contact devices

Military

Mason et al. (2016) analyzed >2700 datasets including Bekker's (1951, 1969) relationship z/d with sinkage over tire section diameter. They proposed a new formula for wheeled vehicles operating on loose dry sands:

$$\Delta CI_{j+1} = \frac{1}{0.2 - 10^{1.65[(\frac{CP}{CI})\,J - 6.0]}} - 4.3$$

where CP is the contact pressure and CI is the cone index or soil strength. This formula will not be easy to use for forestry people.

Footprint forestry machines

The calculations here are empirical and should work year around because a contractor cannot afford to change like the military people change machine or equipment at all cases that are turning up because of tight economic situations and because the terrain could vary on just the single turn.

> *It is recognised that the interaction between an off-road machine and the terrain is complex and difficult to model accurately. Following the empirical approach vehicles are tested in the range of terrain considered to be representative and at the same time the terrain is identified by field observations by simple measurements (citations from Wong, 2010).*

Table 5.2 Compilation of vehicles and ground contact devices tested.

							Load (tonne)				75% of maximum	RTG ground pressure	Rut depth (OL = no load, ML = with load), cm					
Forwarder	Contact	Wheel size	Number of tires	Tire	Tire pressure	Empty	Front	Rear	Load front	Load rear	Front	Rear	1 OL	5 OL	10 OL	1 ML	5 ML	10 ML
Komatsu 860.3	Wheel	710/45 −26.5	8	Trelleborg	6 bar, 3.5; 2.7	19.2	9.8	9	10.5	19.66	54.8	109.9	3.0	5.5	7.2	4.8	9.4	13.9
Komatsu 860.3	Track EcoT	710/45 −26.5	8	Bogie tracks	5	22.4	11.2	10	10.5	20.82	4.5	8.5	0.4	1.8	2.2	1.2	4.3	6.2
Komatsu 860.3	Track EcoOV	710/45 −26.5	8	Bogie tracks	5	22.6	11.3	10	10.5	20.92	4.6	8.5	0.1	1.5	1.9	1.3	3.3	4.0
Komatsu 860.3	Track EcoMagnum	710/45 −26.5	8	Bogie tracks	5	22.7	11.3	10	10.5	20.98	4.1	7.7	1.0	2.4	3.2	1.4	2.9	4.1
Timbear	Track x3		9	Rubber tracks	Steel tyres	15.4	7.2	8.2	9.75	17.94	8.4	8.4	2.4	3.4	3.6	2.7	3.8	4.6
ProSilva	Steel tracks			Steel tracks		23.1	13.0	10	11.25	21.37	20.3	3.3	1.5	2.5	3.0	1.6	2.8	3.9
Rottne 13S	Wheel	710/45 −26.5	8	Trelleborg		25.0	13.0	12	10.5	23.49	26.7	96.6	2.4	4.4	5.3	3.3	6.7	9.4
R13S Harrier	Wheel	710 × 26.5	4	Trelleborg		25.0	13.0	12	10.5	23.49	65.9		–	–	–	1.9	4.4	6.3
Ponsse Wisent	2 + 3 wheel tracks		10	Nokian		20.6	10.7	9.7	9	19.06	10.3	12.8	1.8	3.8	4.9	2.7	6.2	9.4
Ponsse Wisent	Wheel	710/45 −26.5	8	Nokian		16.5	4.6	6.6	9	15.62	19.1	64.2	4.0	6.5	8.9	5.1	9.4	>20
Valmet 860.1	Wheel	710/45 −26.5	8	Nokian	5 bar	17.4	9.9	7.5	9.57	17.07	52	90	3.0	5.3	7.3	1.7	9.7	>15
Elforest	Wheel	600 × 34	6	Nokian	2.4 bar	15.7	5.2	8.7	10.5	19.17	49.8	73.4	3.4	6.7	9.0	5.0	10.0	>15
Valmet 860.1	Tracks	EcoTrack	8	Bogie tracks		19.1	10.5	7.5	9.57	17.07	30.5	68.0	1.3	6.5	10.6	1.6	9.7	>15

According to Wong (1978), ground contact device (running gear) on a vehicle should fulfill the following:

1. Support the weight of the vehicle
2. Cushion the vehicle on uneven terrain
3. Provide sufficient traction for driving and to brake the vehicle
4. Provide adequate vehicle steering control and directional stability.

RTG

In 1969, the Swedish advisory technical group formulated a recommendation (RTG 6904, Anon., 1969) and a model for calculation of a vehicle's ground pressure. For wheels, the formula was **radius x width** (R x W) to calculate the contact area, and for tracks, **1.25 x (radius + length) x width.** The mass is given in kg and the length measures are given in cm. By dividing with 10 (9.81), we get N per cm^2. Later on, this formula was changed to simply **width x contact length** by Malmberg (1981), translated to North America by Joint textbook committee of the paper industry (1989).

> *Ground pressure calculated this way is just a factor among others to judge a vehicle drawbar performance. Ground pressure can only be used with respect to engine power, weight distribution tire patterns and if wheels are pulled or driven (quotation from RTG 6904, 1969).*

The drawback with the old formula was that it was calculated from a sinkage being 15% of the unloaded diameter of the tire (thus ~20 cm), and the elasticity of the soil made it return to half of that value plus 3% increase of the tire width loaded (diagonal tires at that time). Today, when we like to leave no ruts after the operation, this model seems to be somewhat old. The other aspect with too much carbon dioxide going up into the air, the old model by Malmberg (1981) needs to be emphasized: Too much sinkage increases fuel consumption to no practical use (Fig. 5.4).

Comparison between the calculated ground pressures based on the old RTG formula compared (Fig. 5.5) with the measured rut depths after five passages on our organic soil using the old RTG formula for wheels indicated that revision was needed. As seen from Fig. 5.5, the values are in the right direction but precision is missing.

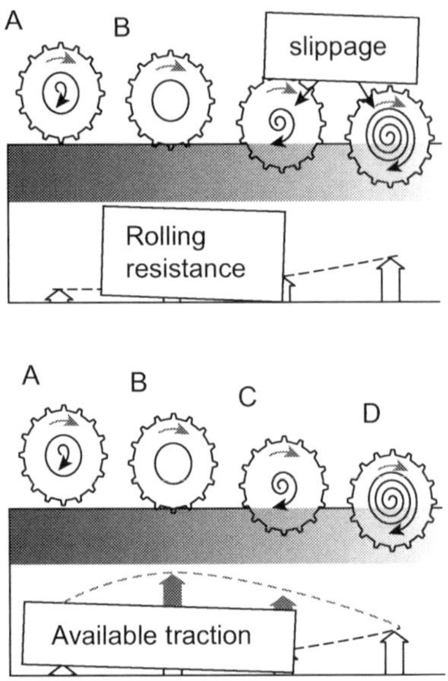

Figure 5.4 Relationship between rolling resistance and available drive force/tractive force. *(After Malmberg, C-E., 1981. Terrängmaskinen del 2, Skogsarbeten, Stockholm. (The Off-road vehicle. Translated part one and published by the Joint textbook committee of the paper industry, 1989, TAPPI, CPPA, Atlanta).)*

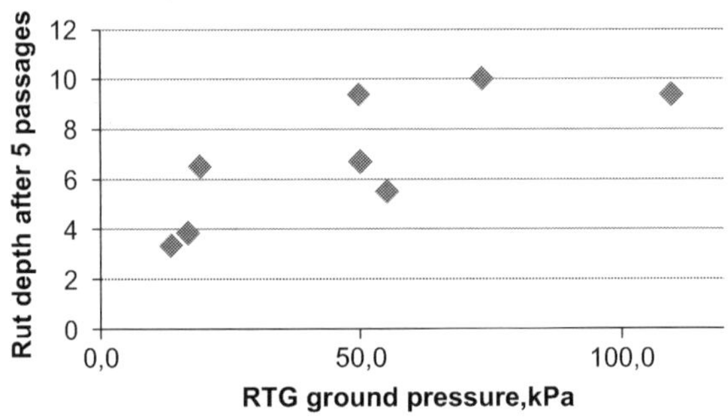

Figure 5.5 Example showing the comparison of measured rut depth driving straight after five passages with calculated RTG ground pressures.

Results

The moisture content in the ground was rather similar between the years (amount of water in gram/wet weight) and was 67% from 2008 to 2009, and from 2010 to 2011, it was 52%–53%, measured on several places and also after few days of testing. Thus, the ruts could be somewhat greater the

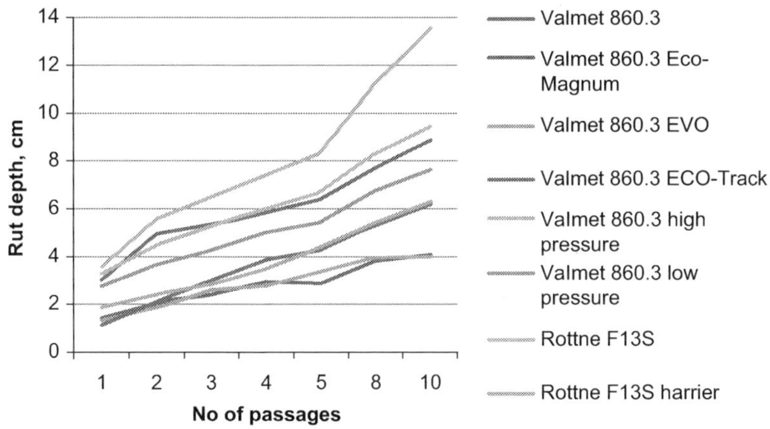

Figure 5.6 Rut development after Rottne F13 S and Valmet 860.3 with different bogie tracks or inflation pressure in the tires, blue = 5 bar, orange = 2.7 bar, or driving with Rottne F13 with the rear wheels running side by side with the front wheels (harrier type of running).

first years, but with experience from that area, it would be just minor effects of the rutting. However, still the surface was so soft that it was impossible to drive with an ordinary car on the area. The other aspect is that it is not like an ordinary forest floor and is missing roots from trees and rhizomes from shrubs frequently growing in the forest floor (Wästerlund, 1989). But there were old roots from wheat and oat plants, although their roots are mainly vertical. The great advantage with this test area is that it is homogenous, which is seldom the case in our forest floor where each single square meter could deviate from the next one in at least one property.

Rutting is shown in Fig. 5.6 as average in left and right side, as well as further on in the text.

Wheels

According to several studies (cf. Karlsson and Myhrman, 1991; Löfgren, 1991; Bradley, 1987), the inflation pressure will influence the contact length since the width of the tire is hardly influenced on a radial or diabelt tire, opposite the older diagonal tires. According to studies made by Wong (2010), the tire will move the soil particles in the surface soil first forward in front of the tire and then backward (Fig. 5.7). This was clearly visible also at our site with high inflation air pressure in the tires (6 bar).

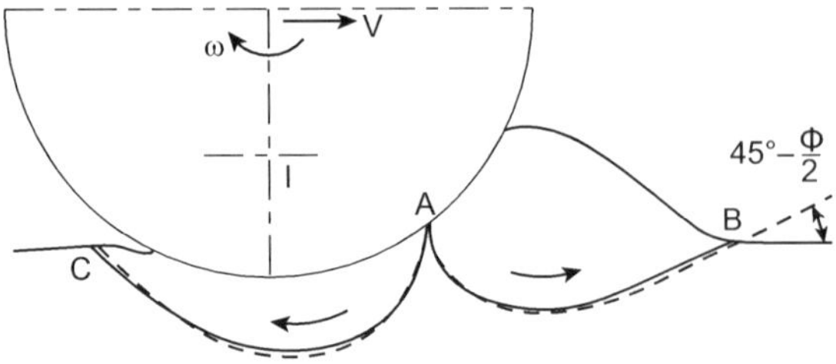

Figure 5.7 In front of a driven tire, there will be a bow wave and then soil particles are pushed backwards with rotating soil particles. *(After Wong, J.Y.,1978. Theory of Ground Vehicles. John Wiley & Sons, New York.)*

Then the underlay has a great influence on how the rut depths are developing, but these experiments were conducted on organic soil to have as homogenous ground as possible and between the years. The chosen experimental area was unusually good in that respect, with a similar moisture content between the years (and previous years), but roots, branches, rhizomes, and grass roots models have from an engineering point of view not been developed.

Proposals for new models for wheels

If you are going to introduce new models, they should not deviate too much from old models to get a general acceptance. But few vehicles have an average ground pressure (dividing sum of contact areas with the total mass of the vehicle). Instead, the load per axle should be controlled (weighted) to be able to point out exceeding load on a certain axle to help to get better weight balance on the whole vehicle (inclusive load). My proposal is to use five different formulas to describe the influence to the ground from the vehicle.

Hegaxy and Sandu (2013) used a number of coefficients to adapt their equations to data having no assortments, which therefore was called numeric. The formulas should have defined assortments, but the idea to use specified coefficients could make the formulas more general with the coefficient that could change with new technical evolutions.

$$\text{Ground pressure } GP_g = W/N \times (D \times B) \times k_g \times 1/1000 \tag{5.1}$$

where GP_g is ground pressure in N/m^2, thus Pa and 1/1000 is KPa, which is the most common unit for ground pressure

W = axle load in N

D = wheel diameter, m

B = wheel width, m

N = number of wheels on one side of the vehicle, thus two for a bogie axle since the rut depths occur on both sides

K_g = wheel coefficient, here simply put to 1 as Wismer and Luth did (Hegaxy and Sandu, 2013)

Spraying color on the hard ground under a 750/45 × 26.5 tire, the contact area became 35 cm long at 5 bar inflation pressure and 11 tonnes load (which is not far from according to Eq. (5.1)).

The rut depths are dependent on soil type, wheel type, and inflation pressure (Löfgren, 1991 and Fig. 5.8) after five passages:

$$\text{Rut depth } S = 2 \text{ X } k_{rd} \times 2 \times GP_g \times n\ 0.1 \times I) \tag{5.2}$$

where N = number of passages

I = inflation pressure in the tire, bar

K_{rd} = rutting coefficient for wheels, here put to 0.01 for up to 10 passages and at more than 11 passages to 0.02.

The subsoil effect (ss) depends mostly on the heaviest axle load, and it seems that the subsoil effect depends mostly on the heaviest axle (cf. Fig. 5.10). For example, Jansson and Johansson (1998) and Eliasson and Wästerlund (2007) measured subsoil effects down to 30 cm depth for a 20 tonnes vehicle and 50 cm depth for a 40 tonnes vehicle. On the other hand,

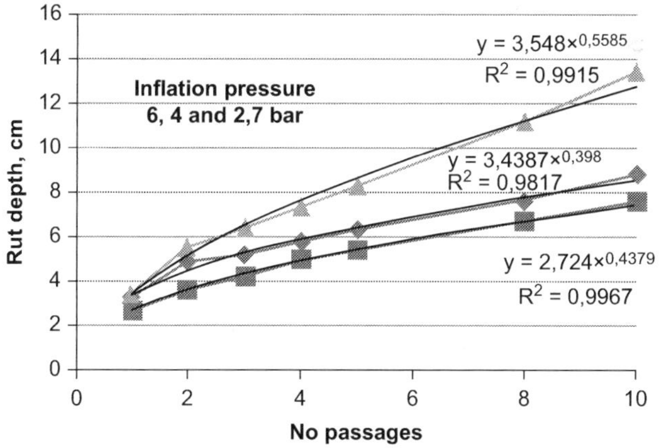

Figure 5.8 Measured rut depth at different inflation pressures measured with Valmet 860.3 with load and number of passages, green (light gray in print version) = 6 bar, blue (gray in print version) = 4 bar, and red (dark gray in print version) = 2.7 bar.

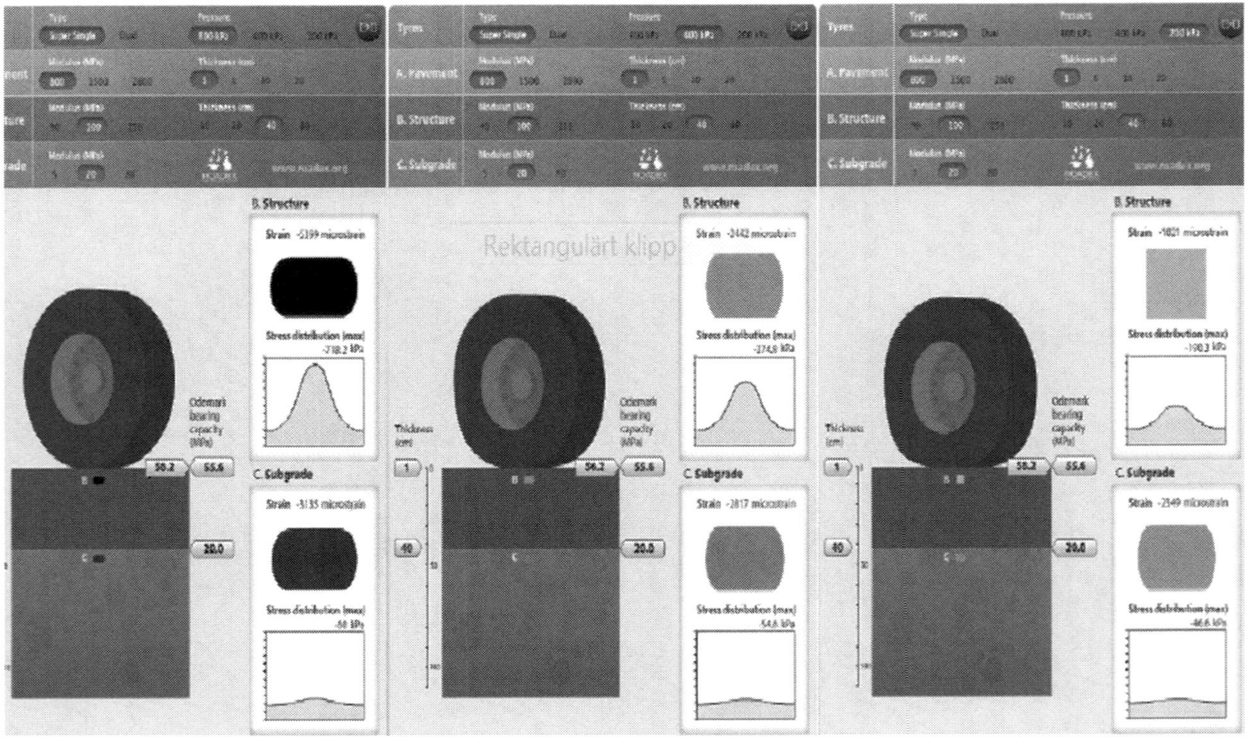

Figure 5.9 Models concerning truck tires (axle load 8 tonnes) and their influence on the subsoil depending on inflation pressure and supersingle tire with 800 to the left and supersingle with 230 KPa to the right RoadEX seminar Elearning (Saarenketo, 2012: Deformation). Observe that the figure down at the bottom represents depth of 100 cm.

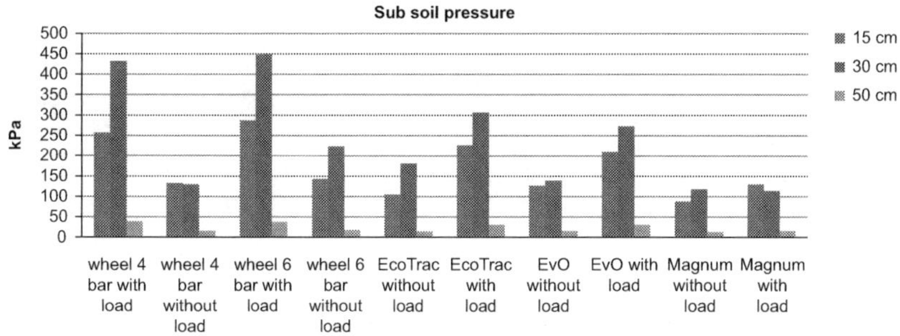

Figure 5.10 Subsoil pressures at 15, 30, and 50 cm depth in the ground for Valmet 860.3 in 2011 test with 20–22 tonnes load on the rear axle and with different inflation pressure and bogie track equipment.

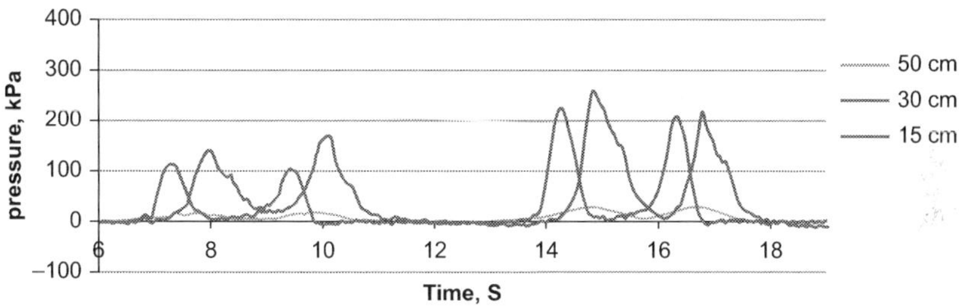

Figure 5.11 Examples of measured subsoil pressures when Valmet 860.3 passed the pressure sensors with EVO bogie tracks traveling to the left. Note that the bogie tracks gave some support between the tires at depth when installed right.

there were no effects on the soil after a 5 tonnes spacing machine. Thus, a proposal

$$\text{Sub soil effect SS} = \text{GPss} \times (W - 5) \times k_{ss} \tag{5.3}$$

where subsoil is then the axle load W in tonnes, depth given in cm, with 5 tonnes axle load being acceptable and k_{ss} is a soil coefficient depending on the soil type. Could be for instance the cone resistance like in the North American studies. Here the value is put to 0.4 for wheels, which fits to the present soil type and the average cone resistance there in MPa.

Proposals for tracks

With respect to the military models for tracks, we keep the equation for tracks.

Traditionally, nominal ground pressure (NGP) has been used as a design parameter for mobility (NGP = load/width x length). Rowland (1972) proposed to replace NGP with MMP (average maximum pressure) for comparisons between tracked vehicles on soft ground. MMP is defined as average value occurring under a road wheel in the track. Rowland further proposed based on a number of empirical tests for rubber tracks and steel tracks on firm ground follow the formula:

$$\mathrm{MMP} = (1,26\ \mathrm{W})/(2NB\sqrt{}2\mathrm{rtp})$$

where W is the mass of the vehicle, N is the number of road, B is the width of the track, r is wheel radius, and rtp is the pitch.

However, the properties of the terrain are not included in the formulas, and Gigler (1992) considered that MMP is strongly influenced by the terrain properties.

$$\text{Ground pressure } \mathrm{GP_t} = \mathrm{k_t} \times \mathrm{W}/(\mathrm{N} \times \mathrm{w} \times \mathrm{L}) \tag{5.4}$$

where w is the width of the tracks in m, L is the contact length in m, which for bogie tracks is roughly the distance between wheel axles, W is axle load in N, and k_t is coefficient for ground pressure, here put to 0.4.

Rut depths for bogie tracks (Fig. 5.11) are of course depending in the same way as wheels on the number of passages (wear), although with the difference that the ruts become less deep and therefore the logarithm of the ground pressure is used:

$$\mathrm{Ruts} = 1 - \log(\mathrm{GPt})\hat{}\log\mathrm{In} \times \mathrm{n}$$

where Ln is the natural logarithm for e and n is the number of passages, see Fig. 5.12.

$$\text{Sub soil effect } \mathrm{SS_t} = \mathrm{GP_t} \times \mathrm{W} - 5) \times \mathrm{k_{tss}} \tag{5.5}$$

where W is the highest axle load in tonnes, and k_{tss} is the soil coefficient depending on soil type. Could be a soil parameter, e.g., cone resistance like in the American models. In the calculations here put to 0.05 also for bogie tracks to fit to earlier studies but also to fit to the soil type used here.

The subsoil effect is an expression for how deep into the soil a certain axle load will act on the subsoil. Observe that the effect is not hard connected to the ground pressure but to the load the subsoil could carry (Wong, 2010, Chapters 2–4). In Fig. 5.9, an example of subsoil changes in a forest road is shown with different inflation pressures in the truck tires (RoadEx, deformation).

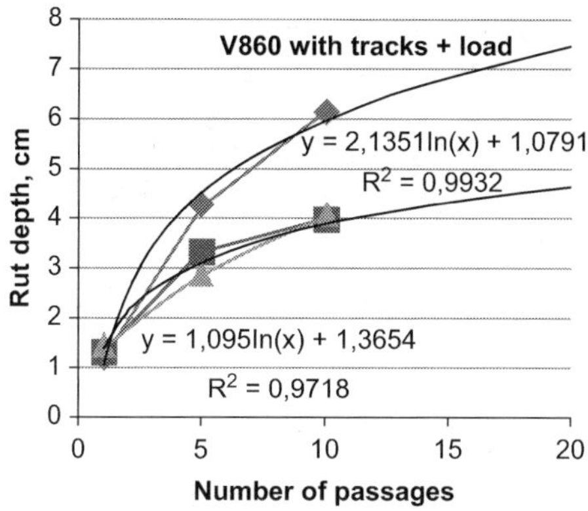

Figure 5.12 Rut formation using Valmet 860.3 with load and three different bogie tracks. Blue (gray in print version) = EcoTrack, red (dark gray in print version) = Evo, green (light gray in print version) = EcoMagnum.

The models here (Fig. 5.13) are used on many forwarders. They are made of steel that has the advantage of increasing the stability of the vehicle in two ways. The weight of the tracks — extra weight near the ground (around 1.5 tonnes per pair) — and with a stiff contact area toward the outer edges compared with an air-filled tire where the elasticity of the tire will make the center of the tire as the tipping point. Furthermore, the contact area will become more rectangular and not oval like a tire, and the rectangular shape spreads out the ground pressure and gives less rutting compared to wheels (Wong, 2010). The disadvantage with bogie tracks is

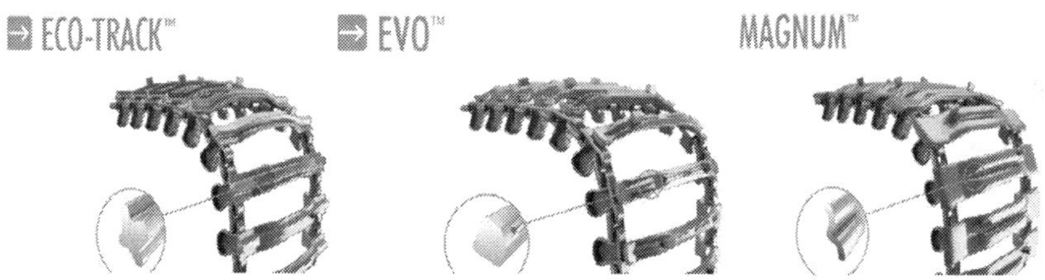

Figure 5.13 Bogie tracks from Olofsfors tested in these experiments. EcoTrack with crossmembers 90 mm wide and link system 20.4 cm, both of which are tractive and supporting, EVO that is tractive with crossmembers 70 mm wide, and supporting Magnum with cross members 150 mm wide. The width of crossmembers is 870 mm on Magnum, while the others have a width of 800 mm.

the extra costs for the tracks. Life length in practical operation may be 5000 to 10,000 h with Olofsfors tracks.

According to subsoil measurements (Fig. 5.11) and the rut formation after driving caused at least 30% rutting, the tracks were not completely tight. Too hard tighten bogie tracks will cause increased rolling resistance. These results are included in the proposed models for tracks as well as the results from the fully tracked ProSilva.

With the cone penetrometer, the soil was tested for variations in the ground. As seen in Fig. 5.14 (to the right), it was clear that there were minor differences across the area. Remarkably, the unloaded Timbear made some soil compaction which could depend on the fact that it had running wheels made of aluminia with a small diameter (60 cm), and the rubber track was stretched around them which made a very short time for soil preparation, and it had a rather heavy front part with the engine. The loaded Timbear had an even weight balance and gave no significant increase of soil density.

Compared with the control, the loaded Valmet 860.3 gave a significant compaction of the soil, which was expected. The proposed models presented here were tested in four different scenarios (Table 5.3).

Discussion

Present studies are done on cultivated peat land, which of cause are missing tree roots and rhizomes, which can increase the strength of the ground with up to 50% (Wästerlund, 1989). On the other hand, it gives the advantage of being rather homogenous ground, which is rather comparable between years and from spot to spot. This could outweigh the drawbacks and make it possible to make a comparative study.

In a number of studies, some kind of penetrometer is used also as an easy measured indicator for the strength of the soil (cf. Anon, 1968). In the big Northern Periphery Project, RoadEx, they used a dynamic cone penetrometer based on a falling weight (8 kg), whereas in North American, a manual cone penetrometer was used. In a ground with logging residues, roots, and rhizomes, few methods with a bigger contact area toward the ground will work properly, and therefore, a simple cone penetrometer will be the best tool to estimate the ground properties.

In the Tierp experiment, it was used as a measure of the strength properties of the ground (roughly half of strength in till soil) but also to give a measure if there was variation of strength in the ground. However, the

Table 5.3 Compilation of used data for the testing, with rutting for the different model Eqs. (5.1), (5.2), and (5.5) and number of passages (1 and 10) + according to the model presented by Hegaxzy and Sandu (2013) H & S.

Bar				Twin forestry 4		Rutting, measured		H & S	GP, kPa	Eq. (5.2) rut depth		cm
Pressure	Tire	Diameter, m	Width, m	Mass, kg	RTG	1	10	Numeric	Eq. (5.1)	1	10	Eq. (5.3) ss
5	710 × 26.5	1.34	0.7	20,000	103.1	4.8	13.5	11.24	103.1	5.16	13.0	31
3	720 × 26.5	1.34	0.7	20,000	103.1	2.8	7.6	11.24	103.1	3.64	8.0	31
5	600 × 22.5	1.16	0.6	19,500	137.4	—	—	8.43	137.4	7.82	16.9	40
5.5	track	1.52	0.8	22,000	88.7	1.7	9.3		35.5	2.0	3.7	30

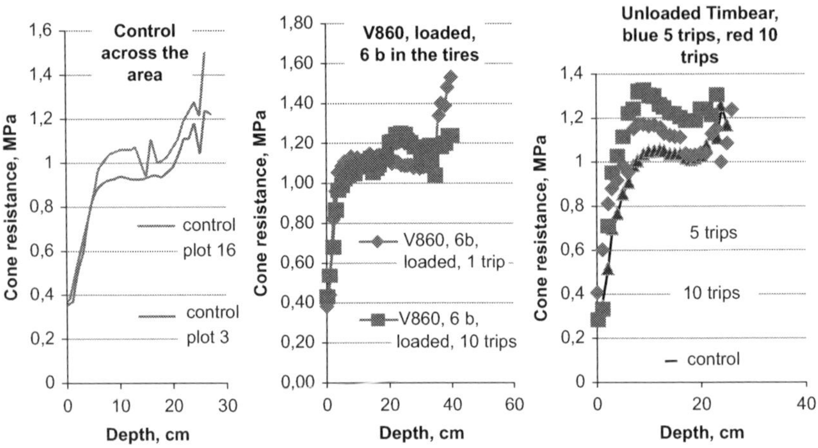

Figure 5.14 Cone resistance values (CI), before driving to the left (2011) across the whole area on two different places, and in the ruttings after unloaded Timbear and unloaded Valmet 860.3, after 1–5 trips (blue: gray in print version) and 10 trips (red: dark gray in print version).

newly introduced Arvidsson method to get subsoil pressures seems to give a fair view of the subsoil pressures when a vehicle is passing on the surface. The disadvantage being that a ditch has to be dug out and therefore few or a single case measurement.

Hallonborg (1996) presented a superellipse for calculation of the contact area under a tire. That model has been in consideration, but being too difficult to use in practise since a number of parameters must be known, for example, soil type and moisture content in the ground.

Already Bekker/Söhne pointed out the advantage to use a long contact area compared to small quadratic contact area, which is most important at low loads (Wong, 1978). Low inflation pressure makes tires to have long contact area (radial tires), which makes it more favorable nowadays to reduce the inflation pressure. However, today with big tires on the forwarders with variable inflation, it would demand more than 1000 hp to feed that pump just to fill up the tires within 30 s!

The compilation here concerns mainly straight on driving, since today the temporary forest roads are rather straight with few and mainly big radius on the curves, which favor bogie constructions and having 8 wheels or 10 wheels on the forwarder (cf. Ponsse Visent with 10-wheel configuration).

Subsoil index cannot be given with a single figure. Here, it is proposed to give it with two figures: ground pressure and subsoil effect. Ground pressure depends not only on the axle load but also on the contact area (wheel size, inflation pressure, and bogie tracks). Subsoil effect depends mainly on axle load according to Söhne's principle studies (see Wong, 1978). Subsoil effect should be less than 20 cm for forest soils.

Practical consequences

1. Look for even weight balance

With these proposed models, it is easy to point out if a vehicle has too uneven weight balance when loaded, which can be done also with the old RTG models. The new models differ rather little from the old ones, the only thing is that it is recommended to calculate the ground pressure per axle and not as average on the whole vehicle. In Fig. 5.15, an example of a loaded 14 tonnes forwarder gains 20% by having a good weight balance compared to a normal version (Fig. 5.15).

We can test the models to get some ideas for the future vehicle and also to test the models. Let us compare a small forwarder with a big one (see Table 5.4):

a. Normal small forwarder, for example, a 12 tonnes Ponsse Visent, 20 tonnes + 12 tonnes load

b. Big forwarder, for example, Valmet V895, 24 tonne + 20 tonne load

- **(1)** Normal situation with load having 60% on rear axle
- **(2)** Balance with equal weight front and rear
- **(3)** Three wheels in the bogie
- **(4)** Bogie tracks
- **(5)** 1—10—20 trips, normal inflation pressure (5 bar) and low pressure for the small forwarder (2.5 bar)

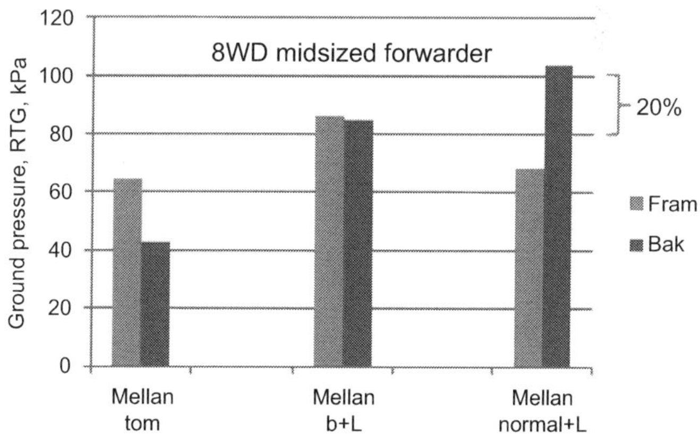

Figure 5.15 Example of what an even weight balance could do for ground pressure calculated according to the old RTG formulas but for single axles.

Table 5.4 Model test of two types of forwarders according to proposed formulas (Eqs. 5.1–5.3 and 5.5).

						Load, MN				Ruts (Eq. 5.2), cm		
										1	10	20
Vehicle	Tire	Diameter, m	Width, m	L, m	Mass, MN	Front	Rear	Eq. (5.1)	Eq. (5.3)	trips		
Small forwarder, 8WD	710 × 26.5	1.34	0.71		305	100	205	105.7	66	5.3	13.3	21.4
Balance	710 × 26.5	1.34	0.71		305	150	155	79.9	34	4.0	10.1	16.2
D:o	Track	—	0.81	1.52	340	170	170	27.1	13	1.1	3.3	3.9
D:o in balance	Track		0.81	1.52	340	110	230	36.7	26	2.0	3.8	4.8
3 wheel	710 × 26.5	1.34	0.71		330	160	170	58.4	28	2.9	7.4	11.8
2 wheel, 2.5 bar, balance	710 × 26.5	1.34	0.71		305	150	155	79.9	34	2.0	5.0	8.1
Large forwarder 8WD	780 × 28.5	1.45	0.78		450	180	270	117.1	103	5.9	14.7	23.7
Balance	780 × 28.5	1.45	0.78		450	220	230	99.7	72	5.0	12.6	20.2
D:o	Track		0.88	1.72	480	230	250	64.8	52	1.8	4.9	6.9
3 wheel	780 × 28.5	1.45	0.78		470	220	250	72.3	58	3.6	13.6	25.1

Figure 5.16 Example of how the subsoil is disturbed due to truck traffic (RoadEx II Deformation).

2. Big or small forwarder

Calculations indicate as expected that a big forwarder makes deeper ruts than a small one and has deeper subsoil effect, although the ground pressures do not indicate big difference. However, if we regard the models as rather true, a transport of 100 m^3 with a big forwarder (5 trips) would give less rutting than eight to nine trips with a small one with traditional unbalance. If the small forwarder has good bogie tracks, it could drive up to 20 times and still make less rutting than one trip with the big forwarder.

Consequences

- Biggest effect is to make the forwarders, so when they are fully loaded, they have a good weight balance (Fig. 5.16).
- More wheels have minor effect on rutting.
- A combination of weight balance and low inflation pressure could be a simple way to modify present forwarders. The drawback could be that rolling resistance could increase the fuel consumption.
- Good type bogie tracks could considerably reduce the rutting.

Acknowledgments

The studies have been made with help of a number of coworkers. Johan Arvidsson, SUAS, and coworkers have borrowed out equipment and several of his coworkers helped us with the measurements. Petrus Jönsson, Martin Englund, Niklas Fogdestam, and Hagos

Lundström at the Forest Research Institute of Sweden and Jeanette Edlund at SLU have contributed to the measurements and are hereby thanked for their good work. The studies were financed by the Forest Research Institute of Sweden under the head of **Björn Löfgren** who was also the leader of the experiments.

References

Alakukku, L., Weisskopf, P., Chamen, W.C.T., Tijink, F.G.J., van der Linden, J.P., Pires, S., Sommer, C., Spoor, G., 2003. Prevention strategies for field traffic-induced subsoil compaction: a review. Part 1. Machine/soil interactions. Soil Tillage Res. 73 (1/2), 145–160.

Anon, 1968. Glossary of terrain-vehicle terms. J. Terramechanics 5 (2), 53–69.

Anon, 1969. Angivande Av Marktryck för Skogsmaskiner. Rådgivande Teknikergruppens Rekommendation Nr 6904. Forskningsstiftelsen Skogsarbeten, Stockholm (in Swedish).

Arvidsson, J., Trautner, A., van den Akker, J.J.H., Schønning, P., 2001. Subsoil compaction caused by heavy sugarbeet harvesters in southern Sweden II: soil displacement during wheeling and model computations of compaction. Soil Till. Res. 20 (1–2), 79–89.

Bekker, M.G., 1951. Soil Vehicle Mobility Report No 19. In: An Introduction to Research on Vehicle Mobility; Part 1: Stability Problem of Soil. US Army Research Laboratory, Aberdeen Proving Ground, MD, p. 233.

Bekker, M.G., 1969. Introduction to Terrain-Vehicle Systems. University of Michigan Press, Ann Arbor; MI.

Bjarnert, J., 2010. Reducering Av Markskador Vid GROT-Skotning. Arbetsrapport 279. SLU, Inst. F Skoglig Resurshushållning, Umeå. [Reducing ground damage during slash – forwarding] (In Swedish with English summary).

Björkhem, U., Fries, J., Lundeberg, G., Scholander, J., 1983. Studier över körskador vid gallring. Inst. Skogsteknik, SLU, Garpenberg. Stencil 253, 73 pp. ISBN 91-576-1778-3. [Studies about soil damage at thinning operations] (In Swedish, with English summary).

Bradley, A.H., 1987. The Effect of Reduced Tire Inflation Pressures on Road Damage: A Literature Review. FERIC, Special Report SR-123.

Brunberg, B., 2004. Produktivitetsnorm för Skotare. Rapport Nr:2004:3. Forskningsstiftelsen Skogforsk, Uppsala [Productivity norm for forwarders] (In Swedish).

Bygdén, G., Eliasson, L., Wästerlund, I., 2003. Rut depth, soil compaction and rolling resistance when using bogie tracks. J. Terramechanics 40 (3), 179–190.

Carter, E.A., McDonald, T.P., Torbert, J.L., Februari 1999. Application of GPS technology to monitor traffic intensity and soil impacts in a forest harvesting operation. In: Proceedings of the 10th Biennal Southern Silvicultural Conference, Shreveport, LA, pp. 16–18.

Edlund, J., Bergsten, U., Löfgren, B., 2012. Effects of two different forwarder steering and transmission drive systems on rut dimensions. J. Terramechanics 49, 291–297.

Eliasson, Wästerlund, I., 2007. Effects of slash reinforcement of strip roads on rutting and soil compaction on moist fine-grained soil. For. Ecol. Manag. 252, 118–123.

Froehlich, H.A., 1989. Soil damage, tree growth, and mechanization of forest operations. In: Proceedings of the Seminar on the Impact of Mechanization of Forest Operations to the Soil. Louvain-La-Neuve, Belgium 11-15 September 1989. Published by the Ministry of Agriculture, Brussels, Belgium, pp. 77–82.

Gigler, J.K., 1992. Design of a Simulation Model for the Prediction of the Ground Pressure Distribution under Tracked Vehicles. Diss. University College, Dublin.

Greacen, E.L., Sands, R., 1980. Compaction of forest soils. A review Australian Journal of Forest Reseaerch 18, 163–189.

Gyldberg, B., 1993. Studies of the Workings of a Powered Disc Trencher. Sveriges Lantbruksuniversitet, Institutionen för Operational Efficiency, Garpenberg. Report No 193. Diss.

Hadas, A., 1994. Soil compaction caused by high axle loads - review of concepts and experimental data. Soil Tillage Res. 29 (2–3), 253–276.

Håkansson, I., 1990. Soil compaction control - objectives, possibilities and prospects. Soil Technol. 3, 231–239.

Hallonberg, U., 1996. Super ellipse as tyre-ground contact area. J. Terramechanics 33 (3), 125–132.

Hegazy, S., Sandu, C., 2013. Experimental investigation of vehicle mobility using a novel wheel mobility number. J. Terramechanics 50 (5–6), 303–310.

Jansson, K.-J., Johansson, J., 1998. Soil changes after traffic with a tracked and a wheeled forest machine: a case study on a silt loam in Sweden. Forestry 71 (1), 57–66.

Jansson, K.-J., Wästerlund, I., 1999. Effect of traffic by lightweight forest machinery on the growth of young Picea abies trees. Scand. J. For. Res. 14 (6), 581–588.

Karlsson, L., Myhrman, D., 1991. Spårdjupsprov 1990. Resultat Nr 5 (1991), Skogsarbeten, Kista [Rutting experiment 1991] (In Swedish).

Klepac, J., Reutebuch, S.E., Rummer, B., 1999. An Assesment of Soil Disturbance from Five Harvesting Intensities. ASAE paper no. 99-5052.

Lull, H.W., 1959. Soil Compaction on Forest and Range Lands. U.S.D.A., Forest Service, Washington (Misc. publ).

Löfgren, B., 1991. Lägre Lufttryck Ger Mindre Spårdjup. Resultat Nr 23. Skogsarbeten, Kista. Less inflation pressure gives less rutting] (In Swedish].

Malmberg, C.-E., 1981. Terrängmaskinen del 2, Skogsarbeten, Stockholm. (The Off-road vehicle. Translated part one and published by the Joint textbook committee of the paper industry, 1989, TAPPI, CPPA, Atlanta).

Mason, G.L., Vahedifard, F., Robinson, J.D., Howard, I.L., Kinley, G.B., Priddy, J.D., 2016. J. Terramechanics 67, 25–36.

Ritari, A., Lähde, E., 1978. Effect of site preparation on physical properties of the soil in a thick-humus spruce stand. Commun. Inst. For. Fenn. 92, 1–37.

Rowland, D., 1972. Tracked vehicle ground pressure and its effect on soft ground performance. In: Proc. of the 4th Int. ISTVS Conf. Stockholm-Kiruna, Sweden. April 24-28.

Saarenketo, T., 2012. Permanent Deformation. Roadex eLearning Package www.RoadEx.org.

Wågberg, C., 2001. Miljöeffekter Och Omfattning Av Spårbildning Vid Slutavverkning. Avd. F. Skogsteknologi, Inst F Skogsskötsel, SLU, Umeå. Examensarbete Studentuppsatser 48 [Environmental impact and extent of log hauling tracks at clear cutting] (In Swedish with English summary).

Wästerlund, I., 1988. Strength components in the forest floor restricting maximum tolerable machine forces. J. Terramechanics 26 (2), 177–182.

Wästerlund, I., 1992. Extent and causes of site damage due to forestry traffic. Scand. J. For. Res. 7, 135–142.

Wästerlund, I., July 1994. Forest responses to soil disturbance due to machine traffic. In: Proc. of Interactive Workshop and Seminar, FORSITRISK, on Soil, Tree, Machine Interactions. ECE/FAO/ILO, IUFRO P3.08-0, EU Concerted Action CEET. Feldafing, Federal Republic of Germany, pp. 4–9, 24 p. (not paginated).

Wong, J.Y., 1978. Theory of Ground Vehicles. John Wiley &Sons, New York.

Wong, J.Y., 2010. Terramechanics and Off-Road Vehicle Engineering. Butterworth-Heinemann, Amsterdam.

CHAPTER 6

Soil mechanics for forestry ground and measurements

Introduction

"Soil is made of nature and not of man. Constructions of bridges and other artificial structures need only knowledge about mechanics... But, as soon we pass from steel to soil, the general theories disappear to exist ..! Within soil mechanics never the precision in calculations exceeds the raw estimations and the principal function in a theory to learn from and observations within the subject area." (Karl Tezaghi in Theoretical Soil Mechanics, 1953).

Terramechanics—what is that?

Common soil mechanics is above all a geotechnical problem for builders of houses or roads as well as constructors of soil working machinery. The strength and deformation are their important questions. For transports and work off-road, the mobility aspect is an important question and the study of soil-vehicle questions started with self-propelled vehicles became more common, thus the beginning of 18th century. The modeling of a vehicle contact with ground became important questions for mobility.

Bekker (1956), Söhne (1969), and Wong (1978) developed theories for off-road locomotion (terramechanics). U.S. Army and German Army and Canadian army development was sponsored by above all the army why the applications were at first for tracked vehicles. Today that development and applications are rather mature. Wheel and tires were more difficult to handle.

During 1970s and 1980s, the focus was turn to "smoother." Soil compaction after vehicular traffic became almost a fashion word and

Soil and Root Damage in Forestry
ISBN 978-0-12-822070-2
https://doi.org/10.1016/B978-0-12-822070-2.00006-7
© 2020 Elsevier Inc.
All rights reserved.

especially within agriculture the result was visible both in the fields and in research activity. Soil mechanics within off-road area became more about "trafficability,"[1] rather than mobility.[1]

The same development in forestry has also occurred although internationally rather small production, except for Scholanderdiss. at SLU in Sweden (1973). Bekker (1980) came to the conclusion after going through the research within forestry in Nordic countries: "*Forestry transportation literature, as related to the environment though unique and important, is underdeveloped—particularly from the viewpoint of vehicle/terrain/machine systems analyses.*"

Thus, let us continue with terramechanical theories and see how they fit to forestry terrain.

What is forestry terrain?

One explanation to why forestry terrain is poorly known when soils are discussed is simply the fact that it is so poorly known in detail.

According to Hägglund and Svensson (1982), in our Scandinavian forestry forest area most common soil type is a till soil and the particle distribution is a sandy—silty till soil type, which means a very wide particle distribution from stone to clay material. The material is created during the glacial period which gives a partially heterogenous material like till soils and partially a very homogenous material as sediments or a peat soil. And the soil type is a Podzol with a humus layer which is 3—10 cm thick with a bleaching layer that is 6—10 cm thick. This means that the soil is both heterogenous and that layers have different properties. The forest floor is often grown with grass, mosses, or shrubs, having tree crowns high above them!

Soil

The forestry ground is usually rather porous and the upper part is easily interwoven with roots, which Sutton (1969) and Scott Russel (1977) pointed out. The number of roots and rhizomes amounting to more than totally 6 cm^2 per m^2 surface area. The soil consists usually of air, water, and particles (see Fig. 6.1).

[1] Definitions according to The International Society for Terrain-Vehicle Systems (ISTVS) glossary (Anon., 1968, 1977): **Soil trafficability** = the capacity of soil to withstand the passage of vehicles. **Mobility** = The ability of a vehicle to reach from one place to another place keeping its ability to maintain its function.

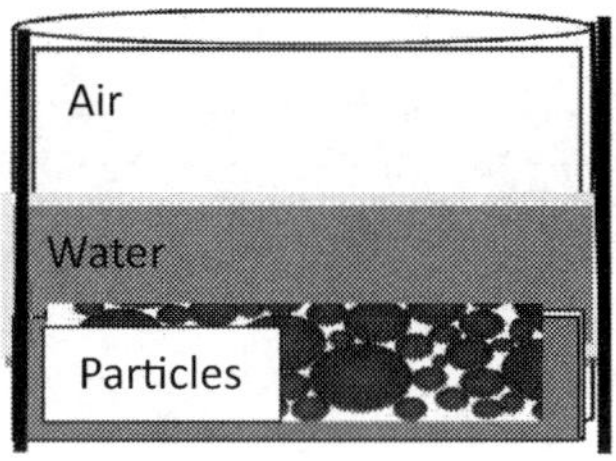

Figure 6.1 Three-phase soil model.

The particles could be classified as friction soil (sand), or as cohesion soil like silt or clay, or organic soil as peat. The particles could be rounded (washed like sediments) or be angular as glacial deposits. The glacial period is giving all kinds of behavior of the particles.

Looking at Fig. 6.2, the principal soil deformation could be *(i)* **plasticity**, which is nondeformable but with constant volume, thus a fluid part of the soil but when drying gives cracks. *(ii)* **Compaction** is a process whereby the soil particles are forced into a closer state of packing with a corresponding reduction in volume and the expulsion of air, with increased

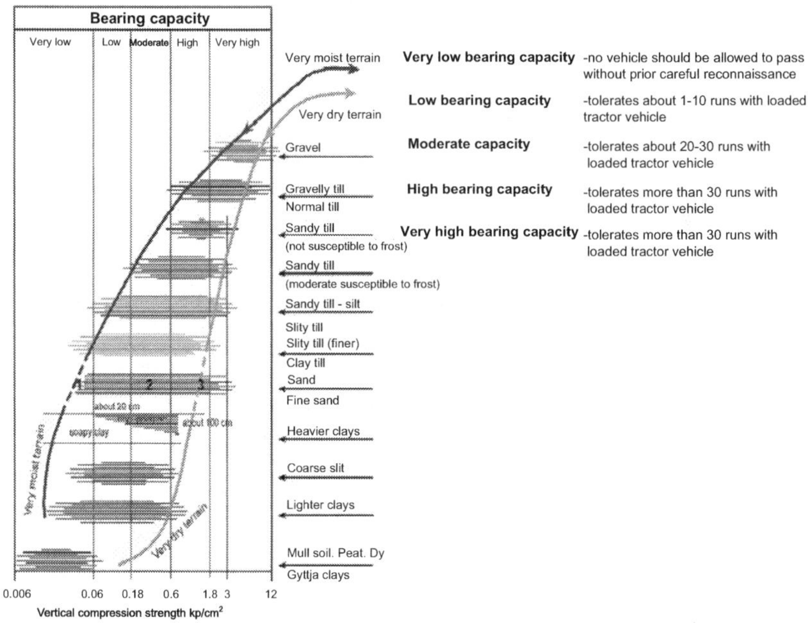

Figure 6.2 Soil strength depending on soil type and water amount in the soil (Swedish Board of Forestry).

load the volume decreases with expel of air. Finally, we have *(iii)* **consolidation,** in saturated cohesive soils, the effect of increasing the load is to squeeze out some of the pore water where water is also pressed out (Whithlow, 1993). The most susceptible soils are normally consolidated clay and silts. All these conditions can exist in a wheel rut, but the effect on vegetation is very different, most detrimental is the consolidation.

Water

Water content in the soil has a great influence how the soil behaves under loading, and the finer the particle size, the more vulnerable the ground is for the bearing strength. At dry conditions, most soils can withstand quite high loads (Fig. 6.3), but when it is wet already a load of 0.2 kp/cm^2 can mean deep rutting.

Principal load reaction is always counter parted with an opposite reaction (normal load), regardless of the size and form of the load. If the load is higher than the strength of the soil withstand, we will get soil compaction.

Measurement methods

Soil penetrometer. It is the most common instrument for testing of soils and frequently used in the army, especially in North America (Fig. 6.4). Mason et al., 2016, did a review of the data and models and algorithms characterizing off-road mobility operating on loose sand. 2737 tests on sand 23 different wheel configurations.

Figure 6.3 A schematic description of a typical forestry ground: Tree plants, shrubs, and grass with boulders (till soil) and roots, humus layer, and bleach soil. *(Drawing by Sigurd Falck (I Wästerlund, unpublished).*

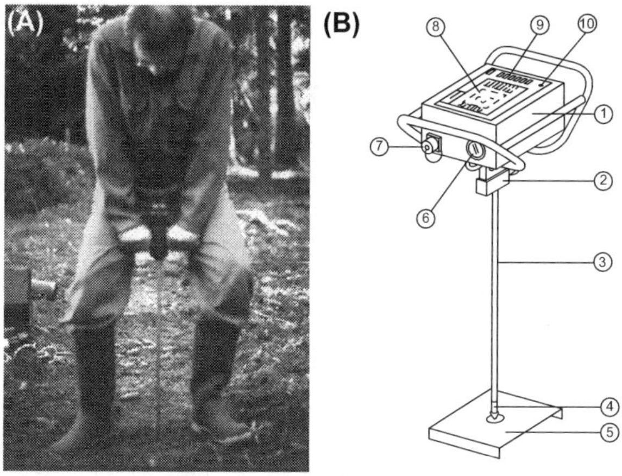

Figure 6.4 A. Handheld cone penetrometer from US Army. B. Soil penetrometer, here an electronic model produced by the Dutch company Eijkelkamp, with automatic reading of soil resistance at determined depth. *(Photo I.Wästerlund).*

The standard cone is 2.1 cm in diameter and basal area is 3.8 cm^2 and 30°cone (ASAE standard) and pressed down with a constant rate of 30 mm-s, it is a rather simple instrument to use but has no precision, especially in forestry ground with roots and rhizomes lying in the humus layer. Splitting a tree root means it is difficult to pull up the cone. A number of things must be fulfilled, first is the size of the cone, which other standard is that it must be pressed down with a constant rate of exactly 3 cm per s, regardless of resistance, because varying rate is measured as the resistance. Thus, a precise rate is very important to get a correct reading. The handheld instrument is tricky to operate and you need three eyes, one to press down the instrument and keep at constant rate, one to check the depth, and a third to read the force. The resistance must be given as kPa and not psi (pounds per square inch). The electronic version is much easier to handle, since it is warning for variations in rate of penetrations (Figs. 6.5–6.11).

Soil cores are also frequently used at soil surveys but tricky to do without disturbing the soil core, especially with roots and rhizomes in the upper part of the soil. Usually 69 mm wide cores are used but roots and rhizomes in the forest soil will make it almost impossible to take undisturbed samples.

GPR. Geo Penetrating Radar is a new technique with promising possibilities but skilled interpreters are needed because you get an image to interpret.

Figure 6.5 Ground-penetrating radar, Malå Geo Science.

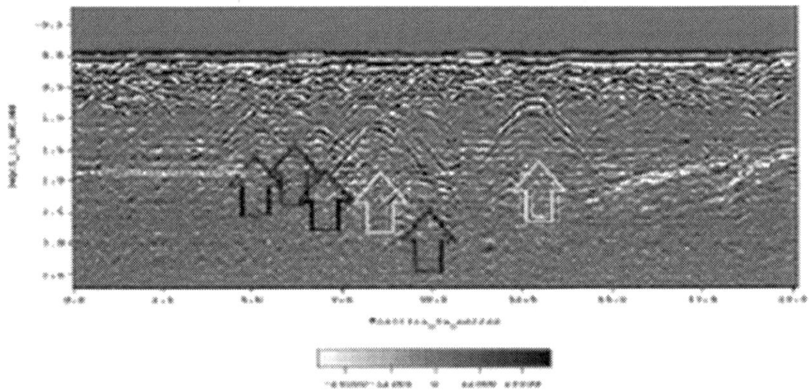

Figure 6.6 A ground-penetrating radargram collected on a historic cemetery in Alabama, USA. Hyperbolic arrivals (*arrows*) indicate the presence of diffractors buried beneath the surface, possibly associated with human burials. Reflections from soil layering are also present (*dashed lines*) *(Wikipedia, 2018).*

The first patent for a system designed to use continuous-wave radar to locate buried objects was submitted by Gotthelf Leimbach and Heinrich Löwy in 1910, 6 years after the first patent for radar itself (patent DE 237 944). A patent for a system using radar pulses rather than a continuous wave was filed in 1926 by Dr. Hülsenbeck (DE 489 434), leading to improved depth resolution. A glacier's depth was measured using ground-penetrating radar in 1929 by W. Stern.

Further developments in the field remained sparse until the 1970s, when military applications began driving research. Commercial applications followed and the first affordable consumer equipment was sold in 1975.

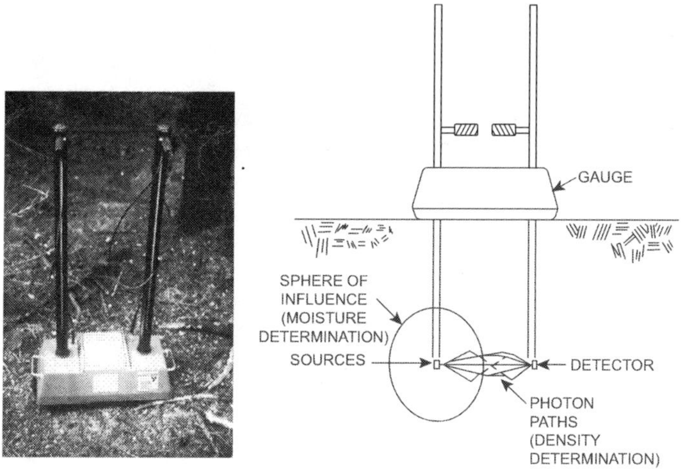

Figure 6.7 Campbell Pacific Nuclear Dual Probe Strata gauge MS-S-24. *(Photo I Wästerlund.)*

Figure 6.8 Simple model of an infiltration tube. Actually, the water level should be in a fixed position. *(photo I. Wästerlund.)*

GPR has many applications in a number of fields. In the earth sciences it is used to study bedrock, soils, groundwater, and ice. It is of some utility in prospecting for gold nuggets and for diamonds in alluvial gravel beds, by finding natural traps in buried stream beds that have the potential for accumulating heavier particles.

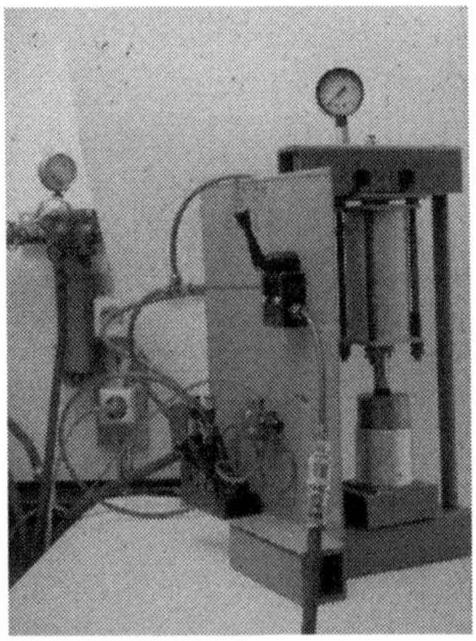

Figure 6.9 Uniaxial compressing equipment, using compressed air and a rubber washer between plunge and soil. *(photo I. Wästerlund.)*

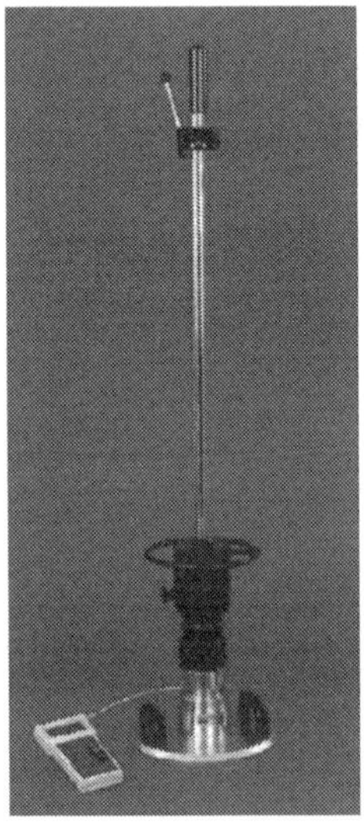

Figure 6.10 Prima, falling weight (FWD) with geophones to test road or pavement constructions. *(photo I. Wästerlund.)*

Engineering applications include nondestructive testing of structures and pavements, locating buried structures and utility lines, and studying soils and bedrock. In environmental remediation, GPR is used to define landfills, contaminant plumes, and other remediation sites, while in archaeology it is used for mapping archaeological features and cemeteries.

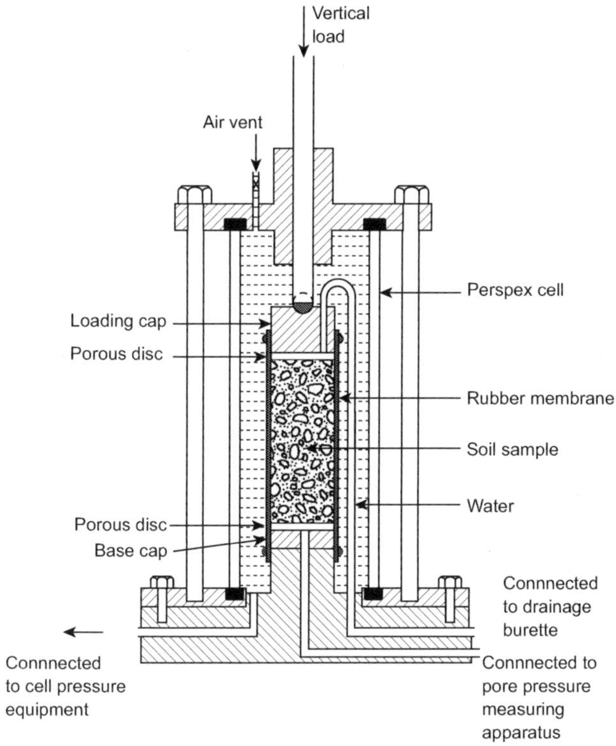

Figure 6.11 Equipment for triaxial measurement of soil samples with pressure from three sides resting on a porous plate. *(from Sutton, 1983.)*

GPR is used in law enforcement for locating clandestine graves and buried evidence. Military uses include detection of mines, unexploded ordnance, and tunnels.

Borehole radars utilizing GPR are used to map the structures from a borehole in underground mining applications. Modern directional borehole radar systems are able to produce three-dimensional images from measurements in a single borehole.

One of the other main applications for ground-penetrating radars is for locating underground utilities. Standard electromagnetic induction utility locating tools require utilities to be conductive. These tools are ineffective for locating plastic conduits or concrete storm and sanitary sewers. Since GPR detects variations in dielectric properties in the subsurface, it can be highly effective for locating nonconductive utilities.

The most significant performance limitation of GPR is in high-conductivity materials such as clay soils and soils that are salt contaminated. Performance is also limited by signal scattering in heterogeneous conditions (e.g., rocky soils) (from Wikipedia 2019).

This equipment can be provided at Malå Geo Science in north of Sweden who have been working with the technology since 1930 and now cooperating with ABEM at Charleston, South Carolina, USA, within the group Guidelines Geo.

Radiation methods

Backscatter method

Make first a hole with the help of guide plate and insert the test probe. MC Portaprobe operates by emitting radiation from two safety-sealed radioactive sources: cesium-137, a gamma emitter for density measurement, and americium-241:beryllium, a neutron emitter for moisture measurement. To determine density, the cesium-137 source emits gamma radiation into the test material. Some of the gamma radiation will pass through the material and be detected by the Geiger—Müller neutron radiation detector located within the unit box. A material with low density will give a high count per time of test. A material of high density will give a low count for the same period of time, as the high density adsorbs more gamma radiation.

To determine moisture content, the americium-241:beryllium source emits neutron radiation into the test material. The high-energy neutrons are moderated by collision with hydrogen atoms. Only on low energy, the materials are detected by the helium-3 detector. A material that is wet will give a low count for the same period of time (from CPN Strata gauge operating manual, 1988). This instrument is mainly used for checking asphalt thickness.

Strata gauge, portaprobe MC-S-24

Both these instruments are supplied with an inbuilt microprocessor. Outside the shipping case (box), no radiation is measured (own tests).

CPN (Campbell pacific nuclear) have two probes working in similar way, but now with two probes (dual probe), with one probe that measures soil compaction(Cs_{137}) down to 600 mm and the other one (Am_{241}:Be) for detection in a small slot in the probe. With this instrument, soil density can be measured in distinct layers. However, the detector in the other probe gives very small signals and is easy to disturb in a moist surrounding. On the other hand, it is easy to take samples by hand to test in the laboratory.

The gamma emitting Cs_{137} is found to be exact determination of bulk density at control measurements (Jansson, 1998), whereas the moisture determinations were less good at high and low values.

This device can actually be used for determination of bulk density also in a stony till soil with Podzol and roots.

Water infiltration

Often soil compaction is associated with reduced pore space and therefore water infiltration can be used to test the rate of water infiltration. The rate, time, and amount water infiltration in wheel ruts compared to control can give a rather fast measure of pore space, size of pores, and orientation. The problem is to install the tube in a forest area due to roots and rhizomes and have no leakage around the bottom walls.

There are a number of tools to determine soil strength in the laboratory. They will be only briefly mentioned and the first two methods give the compression strength of the soil:

- **Uniaxial compression.** Compression of the soil in one direction. For cohesive soils, the diameter of the test cylinder must be at least the same as the height otherwise the wall friction will influence the result.

Proctor Tests

- Proctor test with a falling weight hammering on the sample with, e.g., 25 blows. Gives the maximum density of the soil and the influence of moisture content on maximum density.

The **Proctor compaction test** is a laboratory method of experimentally determining the optimal moisture content at which a given soil type will become most dense and achieve its maximum dry density. The test is named in honor of Ralph R. Proctor, 1933, who showed that the dry density of a soil for a given compactive effort depends on the amount of water the soil contains during soil compaction. His original test is the standard Proctor compaction test; his test was later updated to create the modified Proctor compaction test.

These laboratory tests generally consist of compacting soil at known moisture content into a cylindrical mold of standard dimensions using a compactive effort of controlled magnitude. The soil is usually compacted into the mold to a certain amount of equal layers, each receiving a number of blows from a standard weighted hammer at a specified height. This process is then repeated for various moisture contents and the dry densities are determined for each. The graphical relationship of the dry density to moisture content is then plotted to establish the compaction curve. The maximum dry density is finally obtained from the peak point of the compaction curve and its corresponding moisture content, also known as the optimal moisture content.

The testing described is generally consistent with the American Society for Testing and Materials (ASTM) standards and are similar to the American Association of State Highway and Transportation Officials (AASHTO) standards. Currently, the procedures and equipment details for the standard Proctor compaction test is designated by ASTM D698 and AASHTOT99. Also, the modified Proctor compaction test is designated by ASTM D1557 and AASHTOT180-D.

Other models

A Light Weight Deflectometer (LWD) is a portable falling weight deflectometer used primarily to test *in situ* base and subgrade moduli during construction. LWD measurement is quicker than the isotope measuring method and requires no reference measurements. The equipment has no radioactive sources requiring safety courses and can be operated by one operator, allowing for the analysis of collected data and printing out of data files on site. The most simple LWDs have no load cell but use only a nominal load value, while Dynatest and KUABLWD use a load cell for measurement. The most common LWDs have only one geophone in the center, while Dynatest LWDs can also have sensors at 300 and 600 mm positions. KUABLWDs come with 300 and 600 mm sensors as standard.

A Heavy Weight Deflectometer (HWD) is a falling weight deflectometer that uses higher loads, used primarily for testing airport pavements. Maximum load for HWDs are typically around 300 kN (Dynatest, Carl Bro, and Pave Testing) and 600 kN (KUAB). A common misunderstanding is that a higher load is needed to test an airport's capability to handle heavy aircraft. But, in fact, the testing methods are not designed to test the strength of the construction but to find the material properties.

The test materials are described in ASTM D4694, and the test method is defined in ASTM D4695.140.

Falling weigth deflation

Load impact system

There are two different types of load impact systems: single-mass (e.g., Dynatest, Carl Bro, Pave Testing) and double-mass (KUAB). In a single-mass system, a weight is dropped onto a single buffer connected to a load plate, which rests on the surface being tested. The load force is transferred through the plate, and the plate creates a deflection that simulates a wheel load. In the double-mass system, the weight drops onto a double-buffer system, which includes a first buffer, a second weight, and a second buffer. The principle is based on the law of conservation of momentum with an elastic collision of two unequal masses.

Deflection sensors

Deflection sensors (geophones; force-balance seismometers) mounted radially from the center of the load plate measure the deformation of the pavement in response to the load. Some typical offsets are 0, 200, 300, 450, 600, 900, 1200, and 1500 mm. The deflections measured at these sensors are termed D0, D200, D300, etc. The advantages of seismometers compared to geophones are built-in calibration devices and higher range (5 vs. 2 mm). Geophones are more sensitive to disturbance immediately before the impact since the initial error is integrated. Geophones, however, are much cheaper than seismometers. Dynatest, Carl Bro, Jils, and Pave Testing use geophones, while KUAB have seismometers in their standard FWDs and geophones in their low-cost models.

- **Triaxial strength** test involves also shear strength of the soil by putting a pressure on the soil in three directions.
- **Shear vane.** Shear vane is a rod with wings and by turning it around, the shear strength is measured as the torque needed to turn it around. It does not work on frictional soils (Fig. 6.12). Less good instrument for till soils.

Terramechanics

The general principles are divided between static and dynamic loading.

Static loading

The bearing capacity of a soil is often tested by just loading the soil. By varying the size of the foot, sinkage curves can be made (Fig. 6.13) as well as equations for increase of soil bulk density (Fig. 6.14).

Based on the static load theory, Hadas (1994) made an interesting overview reviewing a number of published articles. His conclusions were

- With a softer soil the distribution of stress goes deeper into the soil.

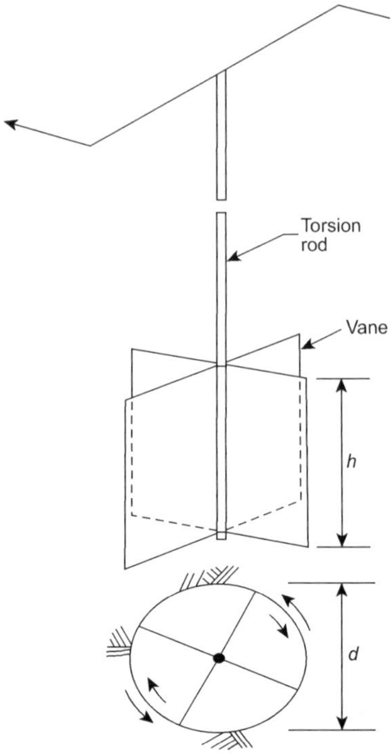

Figure 6.12 Shear vane to measure shear strength, mainly in clay soil. *(from Sutton, 1993.)*

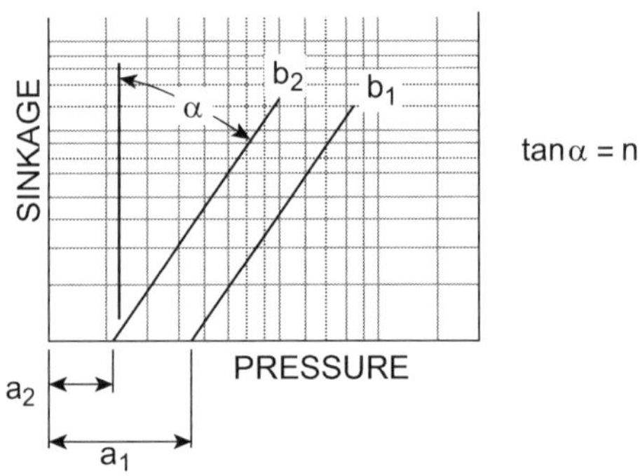

Figure 6.13 Pressure–sinkage relationships. *(after Wong, 1978.)*

- A larger/wider tire dimension will diminish the contact stress.
- Increased axle load will linearly increase stress at certain depth and penetrate deeper. Double the load and the penetration will go 1.4 times deeper.

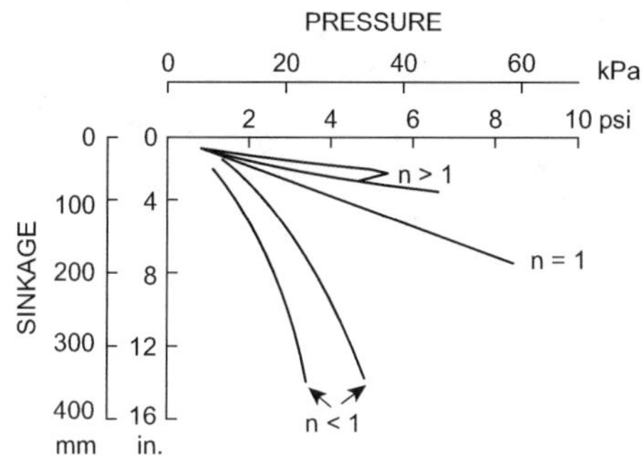

Figure 6.14 With decreasing concentration factor (n) a certain pressure will be transmitted deeper in the soil. *(after Wong, 1978.)*

- Smaller loads than previous should not cause further compaction.
- No substantial differences in soil compaction or soil density between wheeled and tracked tractors may be expected (Fig. 6.15).
- Rate of loading may influence the degree of soil compression (Fig. 6.16).

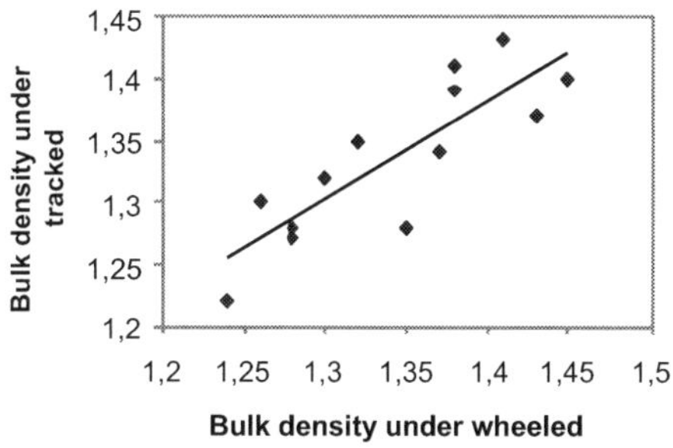

Figure 6.15 Comparison of soil bulk density data after wheeled and tracked tractors.

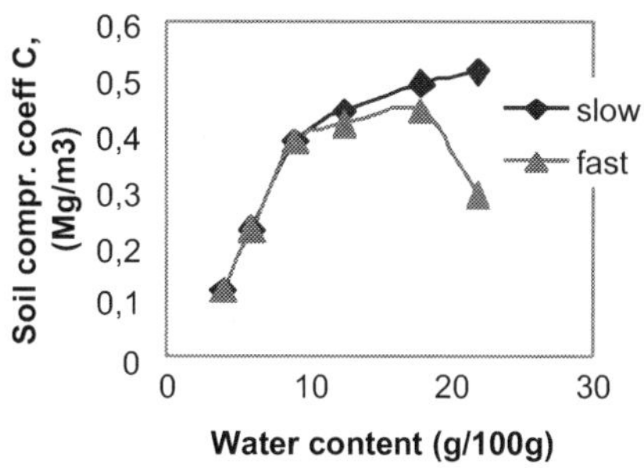

Figure 6.16 Soil compression coefficient C as function of soil water content and compression rate (slow oedometer, fast, Proctor test).

- For a layered soil (soft over hard) it is to be expected a greater stress on top soil.
- It is not the axle load per se that causes most deep compaction, but rather it is the tire dimensions, contact stresses, number of passes, soil density, and water content distributions, the resulting soil strength and the previous stress that causes most deep compaction effects.

Dynamic loading

For field measurements further equipment can be used.

Bekker (1969) introduced the annular shear test (also referred as the Bevameter method). By combining a plate test with a shearing device (Figs. 6.17 and 6.18), the strength of the soil can be determined based on Mohr diagrams. The soil shear strength can be formulated as

$$\tau = c + \sigma \tan \phi$$

where c = cohesion and $\tan\phi$ is the soil friction and σ is the applied stress.

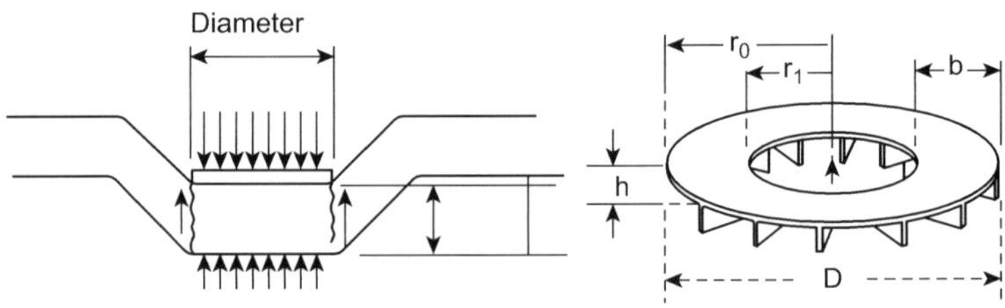

Figure 6.17 Shear annulus device. *(after Wong, 1978).*

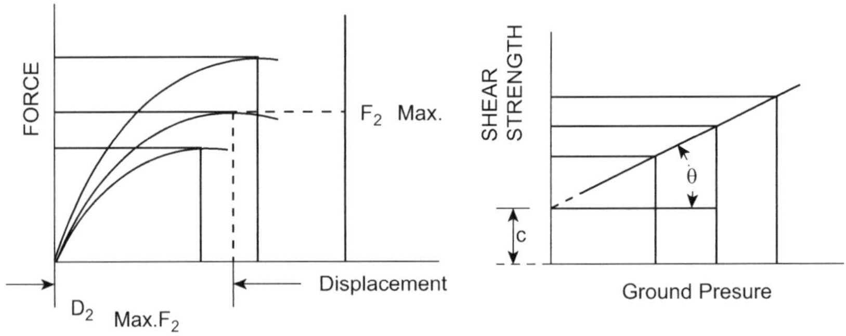

Figure 6.18 Data analysis after shear annulus tests. *(after Wong, 1978.)*

The in situ shear stress tests have mainly been done on nonvegetated areas and may not be the correct method for forest soils with roots. Shoop (1989) tested three methods (direct shear, shear annulus, and triaxial) on calculated data from an instrumented vehicle wheel and found only the undrained triaxial tests related to the performance of the vehicle.

Slip

Most people may think that wheel slip is something very unnecessary and what we refer to may be, e.g., when a wheel is slipping on ice. In terramechanics, slip may be something very necessary to engage the soil strength and get most traction (Fig. 6.19). If the soil is not too brittle (e.g., a dry gravelly soil with round stones), most traction can be obtained at 10%–30% slip. A farmer plowing a clayey field could have his tractor wheels slipping with about 25%–35% to produce the needed traction.

On the other hand, slip may also disturb the soil and introduce shear forces in the soil. Very few data exist on how shear and slip influence soil compaction. One of the few studies (Raghavan et al., 1977) shows that slip may increase soil compaction considerably (Fig. 6.20). Further studies on slip and soil stresses are being done at the National Soil Dynamics Laboratory in Auburn.

Roots

The strength of the tree roots and rhizomes found in the upper horizons must be of importance for the strength of the forest floor. The numbers of

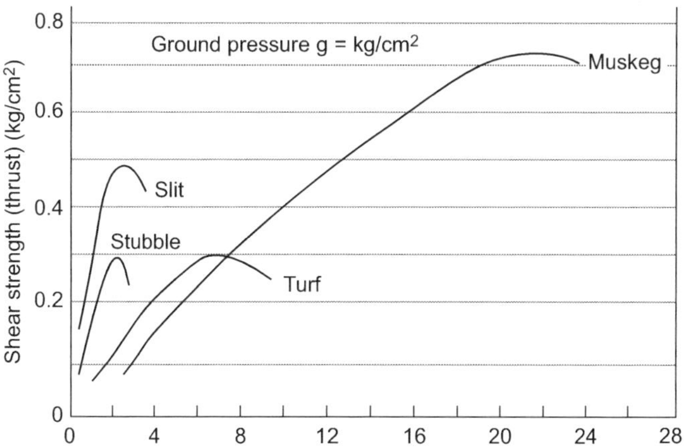

Figure 6.19 Examples of shear strength with varying slip (x-axis).

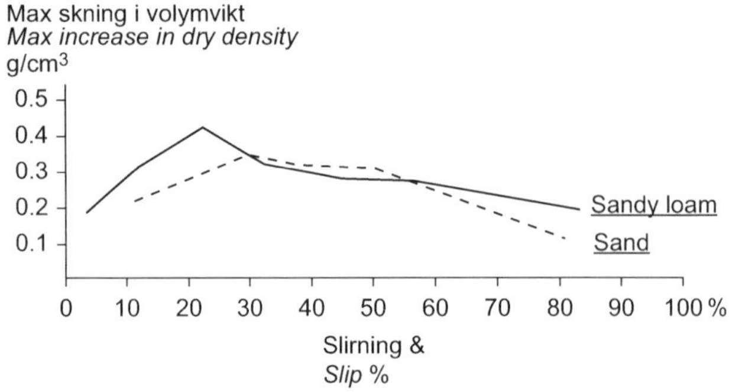

Figure 6.20 Changes in dry bulk density with increased slip. *(redrawn after Raghavan, G.S.V., McKyes, E. & Chassé, M., 1977. Effect of wheel slip on soil compaction. J. agric. Engng Res., 22:79–83.)*

roots and rhizomes may be between 150 and 1000 per square meter surface area. The high extension values indicate that the mat is elastic and do not break until a certain load or shear force is applied on them.

Some models for the root contribution of the soil strength have been published, but they are not easy to use. Wästerlund (1993) proposed a new equation for soil strength calculations with the root component as an additive factor:

$$\tau = c + p \tan + S_r$$

where

τ = soil shear strength, e.g., in N m^{-2}

c = soil cohesion, for coarse grained soils without stones about = 0

p = nominal footprint pressure, kPa

tan = the friction angle, about 0.6 for wheels.

S_r = the root part of the total soil strength, N m^{-2}

$$S_r = k \times RA \times \sigma_t$$

where

k = constant, about 0.5–0.6

RA = tree root and rhizome cross-sectional area in the footprint, cm^2 per m^2 surface area

σ_t = tensile strength of roots, N cm^{-2} cross-sectional area

With 150 cm^2 roots per square meter footprint area, and a root strength of 18–24 kN cm^{-2}, the total soil strength would increase with about 150 kN m^{-2}. This could mean an increase of the soil strength of about

50%–75%. Thus, for both the trafficability and the mobility of forestry machinery, it is important to keep the root mat intact. Observe that this is contradictory to the theories of general soil mechanics.

The forest machine and soil interface

Introduction

Looking at possible machine traffic on the forest floor in a mechanized forestry, we may find up to six machine entries in the stand after the site preparation (Wästerlund, 1992). It is then assumed that the site preparation is successfully remolding the soil damage after the clear-felling operation and a planting machine is attached to the site preparation machine or nonexisting.

The first machine entry is the cleaning/spacing machine which may influence an area of about 20% when doing selective spacing. With row cleaning it may influence up to 50% of the area. One or two entries of that machine may be necessary. Commercial thinning operations could be done 1–5 times during the rotation of the stand. At selective thinning and with road spacing from 15 to 30 m, the influenced area could be 12%–25%. At row thinning the influenced area could amount to 37%. Very few data exist on influenced area at clear-felling from Europe but rough estimations indicate about 30% of the area may be influenced. Deep rut formation and soil damage are reported on 30%–50% on the used tracks but reports on even 100% soil disturbance may be found. Thus, the machine trails from different forestry operations could usurp 15%–35% of the site area and half of that area could be markedly disturbed. These figures motivate clearly ambitions to reduce soil damage from machine traffic.

In a theoretical approach on literature data, Wästerlund (1992) made calculations of probable real ground pressures and tractive forces in the contact area. The calculations were based on results from an Australian study showing that the maximal contact pressures are about 1.6 times the mean contact pressures and acting on the inner half of the contact area radius (= 1/3 of the contact area, Dexter et al., 1988). The skidder with load was assumed to have the weight as the forwarder just to make comparisons between these two types of yarding machinery. The results (Table 6.1) show that the tractive forces are very high in the skidder due to the dragging resistance and a harvester just propelling itself should impose very small shear forces on the soil. Thus, from a soil disturbance point of

Table 6.1 Estimated traction data for three types of forestry machines (Wästerlund, 1992).

Type	Harvester	Forwarder	Skidder
Mass, kg, with load	10,000	18,000	18,000
No. of wheels	4	8	4
Real mean ground pressure ($_m$), kPa	83	115	150
Contact area (A), cm^2	3000	2000	3000
One-third of contact area (A/3), cm^2 *	1000	667	1000
Needed tractive force with uneven torque distribution, kN	2.5	5.3	13.0
D:o for one-third of contact area, N cm^{-2}	2.5	7.9	13.0
Maximum ground pressure, 1.6 $\times_m$, kPa	133	184	240
Extractable traction, $\mu \times_m$, $\mu = 0.6$, N cm^{-2}	8.0	11.0	14.4

*Dexter, A.R., Horn, R., Halloway, R. Jacobsen, B.F., 1988. Pressure transmission beneath wheels in soil on the Eyre peninsula of South Australia. J. Terramechanis Vol.25(2): 135–147.

view, it should be better to carry the timber than to drag it, and harvesters should, with reasonable weights and ground pressures, cause only small damage to tree roots and soil.

The forestry machine

Machine mass and ground pressure

Two Swedish projects have been dealing with the design of the machine and how it influences the soil measured only as rut formation:

1. With increasing number of passages, the rut depths increased and more rapidly if the machine was heavy and/or had a high nominal ground pressure.
2. Generally, the rut depths increased with a combination of machine mass and ground pressure. Some machines had less rut depths than expected without the real reason could be pin-pointed.
3. By increasing the size of the tires from 600 mm wide to 800 mm, the rut depth was reduced to 50%. However, the result is based only on one soil type and should be repeated on other sites for validation.
4. Reducing the inflation pressure from 340 to 100 kPa may reduce more than half of the rut depths. The results indicate also that with high inflation pressure, it may be of benefit to reduce the load somewhat.
5. Four wheeled harvesters had about the same rut depths when driving in a curve as when driving straightforward. Six or eight wheeled machines (with one or two bogies) had doubled ruts depths when driving in a

curve compared to driving straightforward. An 8-wheeled machine with steering of the wheels in the bogie had only a 25% increase in the rut depths when driving in a curve that had even a smaller radius than other bogie equipped machines.

6. In practice, a 600 mm tire with CTI was equal to a 800 mm wide tire without inflation control.

Thus, about equal weight balance or slighter heavier in the rear would be favorable with load.

Transmission

Presently, there are four types of transmissions common on forestry machines. The mechanical transmission has a rigid connection to the engine via gears and axles and is the most energy-efficient type if only considering the efficiency within the transmission. A hydrodynamic transmission with a converter will help the operator to perform on slopes but has some drawbacks in downhill driving. With a hydraulic pump and motor between engine and gearbox, it becomes easier to keep a desired speed. In forestry, there is a great need to have all wheel drive. Since all of these transmission types have fixed gear ratio between front and rear, different loading of the machine will introduce necessary slip in one or the other way. An all-hydrostatic transmission with hydraulic motors in all wheels may have quite low energy efficiency and may cause very little wheel slip. However, since the motors are more or less independent, a slip control may be needed and with a such one, the transmission can be rather good to bring efficient traction to the ground (Olsen and Wästerlund, 1991).

In a Swedish study, it was found that a mechanical transmission with all wheel drive may cause wheel slip and restraints in the transmission just because of different rolling radius of the wheels (e.g., due to different loads). It was therefore of interest to study the torque distribution of an all-hydrostatic driven forwarder (Mohr and Eriksson, 1990). This forwarder had a much better torque distribution than the one with mechanical transmission but still when driving in curves, negative torques could appear on some wheels. An antislip control was developed on the vehicle and different types of control designs were tested. With all wheels separately controlled, slip was almost totally eliminated. With a simpler, patented, and rather cheap solution, a reasonable reduction of induced slip was achieved. At a forced slip situation, slip was reduced with 30% and mobility actually increased. However, slip control seemed not to influence fuel consumption.

Conclusions: Mass, weight balance, ground pressure, and torque distribution are factors that will influence the rut formation. When driving in a curve, four-wheeled machines make less ruts than machines with bogies. However, with a steered wheel in the bogie, skid steering can be avoided and the reduced ground pressure with a bogie utilized.

Slip and tires

Tires can be bias-ply, radial, or a combination. The construction is of importance how the tire behaves in different situations. Reduced inflation pressure can be dangerous for bias-ply tire. Lugs on the tires are to improve the traction, and it seems that some think that high lugs give good traction. It may be so in a few situations, but generally 2–3 cm lugs are enough because higher lugs give higher rolling resistance and net traction may not be improved (Dwyer, 1983).

With a mechanical transmission, the engineers may have built in some differences in the gear ratios between front and rear part. One reason is to give the machine better performance when loaded, and another reason is to give an articulated steered machine better stability. Thus, to start with, we may have 2%–3% slip built-in in the machine. Next thing is that the tires are made with a certain tolerance and usually the diameter tolerance is 1.5%. Assume a tire with 1.1 m in diameter. This tire can have a circumference between 3.4 and 3.5 m giving 2.9% slip if both are running on the same gear ratio.

Slip is often defined as Slip = (1-real trav. distance/theoretical distance) 100, %. With only little traction to propel, our forestry machine forwards, the machine may have about 5% slip.

According to the results, wheel slip plays a greater role in the rut formation than wheel load or ground pressure. At 12% slip hardly any part of the track had damaged surface and only a depression was visible after five runs but at 64% slip the damage was 100% already at the first run.

Conclusions: Slip may increase the wear of the soil and thus increase the rut formation. Driving on only gently undulating terrain and carrying the load the slip is rather small if not the surface structure is rough. But with bogie tracks, we can get a better support for the tires and with right type of tracks it can even take less fuel, thus, environmentally and economic beneficial (Bygde'n and Wästerlund, 2016).

Contact forces

Bogie and tracks

A bogie on a forestry machine may be different to a bogie on a truck, where two axles close enough could be considered as a bogie. On forestry machines, the wheels are often connected in a bogie beam to give both traction and the same speed to both of the wheels. A tandem construction of this latter type gives the wheels the properties of one large wheel that incorporates both of them. Thus, the tandem construction will get good traction and bearing capacity. The bogie beam will also give better riding comfort by decreasing the obstacle heights. However, there are two drawbacks with a bogie. One is the skid steering in curves and the other one is the tendency to lift up the front wheel at high traction. Unless a steering of the wheels in the bogie, the only way to reduce the shear forces is to have tires with round shoulders. The momentum to lift the bogie at high traction depends on the momentum arm, and the new constructions with a small arm has made them almost neutral.

Tracks can be put on a bogie to improve traction and somewhat the bearing capacity and can be of many different kinds. The main thing to observe is how the tracks are constructed. Older types have the linkage between the track shoes somewhat below on the tire to keep track on place on the tire. The problem is that this connection point has a less rolling radius than the tire. The track will run with different speed than the tire and there must be some slip in between. Tests on newer types of tracks have shown less fuel consumption of the vehicle indicating energy was spent on something else than forward movement. For snowy or wet conditions, a good track could improve traction and reduce rutting.

Load transfer to the soil

Quite many machine manufacturers nowadays accept HVO 100 (which is Hydro Vegetative Oils made from butcher waste or nowadays logging residues) and can replace diesel in many engines. Having both a machine that makes no ruts and also running on biofuel we have the machine for both fields and forest.

Unnecessary slip is just a waste of energy just pushing soil from one side to another. By supplying the machine with bigger tires and less inflation pressure and in the same moment, we reduce slip and having lighter vehicles, we may also reduce soil damage. To be able to reduce slip both machine constructors and machine operators should be involved. The first category to make the machines more efficient and the other to handle the machines in a more gentle way.

Conclusions: Slip may increase the wear of the soil and thus increase the rut formation. By driving on only gently undulating terrain and carrying a lighter load the slip is rather small if not the surface structure is tough. But with bogie tracks, we can get a better support for the tires and with right type of tracks it can even take less fuel, thus, environmentally and economic beneficial (Bygde'n and Wästerlund, 2006).

Don't crush the roots!

References

Anon, 1977. ISTVS standards. J. Terramechanics 14 (3), 153–182.

Bekker, M.G., 1969. Introduction to Terrain-Vehicle Systems. Mich. University of Michigan Press, Ann Arbor.

Bygde'n, G., Wästerlund, I., 2016. Tracks in the Forest. OmnSscriptum, Saarbrücken, Germany.

Dwyer, M.J., 1983. The tractive performance of wheeled vehicles. In: Proc. 2nd European ISTVS Conference, 3–5 October, Ferrara, Italy, pp. 35–59.

Dexter, A.R., Horn, R., Halloway, R., Jacobsen, B.F., 1988. Pressure transmission beneath wheels in soil on the Eyre peninsula of South Australia. J. Terramechanis 25 (2), 135–147.

Hadas, A., 1994. Soil compaction caused by high axle loads – review of concepts and experimental data. Soil Tillage Res. 29, 253–276.

Jansson, K.-J., 1998. Effects of machinery traffic in forestry on soil properties and tree growth. Acta Universitatis Agriculurae Suecicae. Uppsala. Silvestria 66.

Mohr, E., Eriksson, U., 1990. Traction forces and torque distribution on an 8WD forwarder. In: Proc. 11th International Conference of the ISTVS, Lake Tahoe, Nevada, USA, vol. 1, pp. 803–812.

Raghavan, G.S.V., McKyes, E., Chassé, M., 1977. Effect of wheel slip on soil compaction. J. agric. Engng Res. 22, 79–83.

Shoop, S.A., 1989. Thawing Soil Strength Measurements for Predictingvehicle Performance. North American Meeting of ISTVS, Victoria, B.C., Canada, pp. 165–182.

Sutton, R.F., 1969. Form and Development of Conifer Root Systems. Commonweath Forestry Bureau, Oxford, Engl, p. 131. Techn. Commun., No 7.

Söhne, W., 1969. Agricultural engineering and terramechanics. J. Terramechanics 6 (4).

Whithlow, R., 1993. Basic Soil Mechanics, second ed. Longman Scientific & Technical, Harlow, New York.

Wästerlund, I., 1992. Extent and causes of site damage due to forestry traffic. Scand. J. For. Res. 7, 135–142.

Further reading

Ala-Ilomäki, J., Saarilahti, M., 1990. Rut formation on peat soil - experiences from a forced-slip wheel tester. In: Proc. 10th International Conference of the ISTVS, Kobe, Japan, August 20–24, 1990, pp. 457–466.

Björkhem, U., Fries, J., Lundeberg, G., Scholander, J., 1983. Studier Över Körskador Vid Gallring. Sveriges Lantbruksuniversitet [Studies of Operational Damages at Thinning]. Dept. Of Operational Efficiency, SLU, Garpenberg. Stencil nr 253.

Jansson, K.-J., Johansson, J., 1998. Soil changes after traffic with a tracked and a wheeled forest machine: a case study on a silt loam in Sweden. Forestry 71 (1), 57–66.

Sutton, B.H.C., 1986. Solving Problems in Soil Mechanics. Longman Scientific & Technical, John Wiley & Sons, New York.

Terzaghi, K., 1953. Theoretical Soil Mechanics. Wiley, NewYork.

Wästerlund, I., 1990a. Soil strength in forestry measured with a new kind of test rig. In: Proc. 10th International Conference of the ISTVS, Kobe, Japan, August 20–24, 1990, pp. 73–82.

Wästerlund, I., 1990b. Damage to the Ground and the Stand after Mechanized Cleaning. Sveriges Lantbruksuniversitet, Inst F. Skogsteknik, Garpenberg. Uppsatser och Resultat Nr 193.

Wästerlund, I., 2017. Biofuels as Liquid or Gas for Engines. Forest Research and Engineering: International Journal. Med Crave Publishing.

CHAPTER 7

Alpine routes reinforced with geonet for ATV

Introduction

Peatlands have low bearing capacity for both walking and terrain vehicles, but on top also very low capacity to recover especially in alpine environment (Fig. 7.1).

Air borne photos with infrared colors in the alpine landscape (Allard et al., 2004) have shown areas with huge damages on ground and vegetation. To some extent, it is depending on the Laplanders (Sami people living in the Sapmi area) work with reindeers, e.g., at calf marking, but some is depending on tourism traffic, hunters, and fishermen. There is a law forbidding private terrain traffic (excluding the Laplanders and forestry work), but it is difficult to control that traffic. On the other hand, just walking can by trampling destroy the vegetation and create rutting in the water-filled peatlands, like on the Kings route (a walking trail) with maybe 1000 persons walking near the Norwegian border but also on the Swedish side there are partially big damages on the ground in the weak ground (Wästerlund, 2006). County ministers are by law forced to repair and channelize the traffic but do not know how. Similar damages are also seen in Alaska, and they try just to fix the worst damages by spade.

The project started with a Norwegian patent (Norvia AS) as a base where a net was just put out on the ground and anchored with big screw types made of plastics. Then ATVs (All Terrain Vehicles) were supposed to be able to travel on that reinforced ground (Fig. 7.2).

AIM

We (Mårten Jonsson and the author of this report) were missioned to

- Design both methods and techniques to do this reinforcement of routes in the alpine environment,
- And to evaluate the method of reinforcement,
- And suggest techniques and recommendations for this work.

Soil and Root Damage in Forestry
ISBN 978-0-12-822070-2
https://doi.org/10.1016/B978-0-12-822070-2.00007-9
© 2020 Elsevier Inc.
All rights reserved.

Figure 7.1 The project leader stuck with his ATV in a mud hole caused by a dug ditch in the alpine landscape and calling for help.

Figure 7.2 Alpine landscape with sparse birch forest growing on top of a peaty landscape. Just water-filled underground peat with no strength but looks ok from above view. *(Photo I. Wästerlund: Color photo.)*

ATVs

The test vehicle we used was an ATV, which are a 4-wheeled motorcycle with a 20–45 hp engine, often one cylinder engine. Most often with a very simple transmission to get 4WD. Thus, it is very dependent on the weight balance, since all wheels are propelled with the same rotational speed.

With aggressive tires, it can create deep ruts, but used most frequently by the Sami people to be able to follow the reindeer herd. The weight (mass) of the vehicle with one driver (weighing 75 kg) was 400–670 kg. The heavier one was a 6WD to carry equipment to the mountain areas. The normal test vehicle was a four-wheeled one, with a contact area under the tires 616 cm^2 front and 560 rear with inflation pressure 50 kPa front and 40 rear.

Geonet

To use a net made from steel or iron or plastic, material like polyethylene (PE) or polypropylene (PP) is actually a rather old product and has been used in the last 20–30 years. Before the plastic geonets time among other things (like wood planks), sheep metal nets were used among other things to stabilize slopes from erosion (see Whitlow, 1990). The technique to use geotextiles fabric was developed as reinforcement of the underground to spread out the load on soft terrain. When better plastic nets could be made to improve the road strength, the overbuilding could be reduced (Anon, 1999). With a geonet with coarse macadam packed in the net, the overbuilding can be halved and thus the cost for the geonet is saved.

The modern types of geonets in thermoplastics could be either of two types: **Thermoplastics** are built from linear or branched polymers bind together with secondary bindings (van der Waals connections, dipole connections, or anion bindings). These thermoplastics become easy to melt and are not suited for reuse. **Polyethylene** (PE) is a plastic with little branching and depending on molecules to coproduce and thereby density as (LDPE) low, between (HPDE), or high (HDPE) density. In general, PE plastics have high strength for punching but are dependent on temperature and the material could shrink when cold. It is very common as water tubes.

Propene (PP) is the most common thermoplastics and is built from propene and therefore often called wrongly polypropylene. The glass temperature is between −10 and −20°C, which means that in the cold winters in north Sweden, it becomes brittle and easy to break. It has a low weight, but high strength and can be colored. It will be easily decomposed by UV light, when not treated with smoke-black (2%). It is used in corrosive environments and can withstand most acids or basic environment. The brittle properties at cold could be reduced with additives or copolymerization with proper material.

Some geonets PP have been developed by making the production in a different way by pulling out and pressing out the material at extrudation giving a very high strength of the material and plane stiff net (Tenax LBO): 33 kN m^{-1} or 44 kN m^{-1}.

Material and methods

Laboratory tests

To study the geonet properties, the tensile strength were tested with a Hounsfield tensile tester where both the tensile strength and elongation were studied. The pulling rate was 50 mm min^{-1}, which is much higher than the standard (1.25 mm min^{-1} according to ASTM). Most of the tested materials had fact sheets but was missing for Norvia AS, but the data sheets did not tell how they break (sudden or tough) and interesting was also how they behave at much higher pulling rate. Together more than a 40 tests were done and three of the nets were of the type plane stiff PP nets. The increased rate did not affect the tensile strength but the elongation, which decreased.

Field tests

Vegetation measurements were done by putting a wooden frame (0.5 m × 0.5 m) on random places along the traveled passages on geonet (more than three per vegetation types) and similarly from untouched ground (at least six per place). The biomass was cut with a scissor near the ground and weighted wet. Theory: Show the amount of vegetation, but we are not specialist to tell what species are growing there, just collect the amount of above ground vegetation.

Plate testing/plate penetrometer was used to measure the strength of the ground with and without armoring (Sirois et al., 1989). With the registration penetrometer (Eijkelkamp penetrologger), a round plate screwed on the penetrometer (10 cm^2, radius 17.85 mm) was pressed down by hand with roughly 3 cm s^{-1} and the depth measured ultrasonic. The force to press down the plate was max 500 N (70% of the mass of the handling person). The plate had to be of that size to fit over squares in the net.

The new constructed plate was 13 cm × 7 cm (=91 cm^2). The loading was applied with a hydraulic cylinder powered by a small hydraulic pump powered by a small petrol engine, pressure and sinkage were measured once per second with help of a geared potentiometer.

The shear strength between the ground/geonet and the wheels should be done dynamically, but the breaks of the vehicles were too bad so we had to do it static out in the alpine area (Wong, 1978). To measure the rolling resistance, we pulled one ATV using the electric winch on the 6WD and pulled with a rate of 0.17 m s^{-1}.

The tensile tester was a Bofors KRG-4 load cell, which was rated to 5 tonnes as maximum pulling strength.

Anchoring was tested on five different types of anchoring. They were put in for 2–3 h before testing and was pulled up at 20 degrees angle simulating a person pulling up the geonet (a trailer with a big tube to roll up/off the geonet) A string was put around the top of the anchoring plug and pulled up with a rate of 0.17 m s^{-1} with the winch on the ATV.

The 6WD had a rubber track around the rear bogie wheels, which also was the vehicle used for putting out (rolling) the geonets. It was often raining during our measurement and Fig. 7.3 can show the water puddles during the tensile testing on the geonet put out the year before side by side to the old rutted trail (on the right side) and water running across the route.

Results

The tests with the plate showed and the tensile tester showed that many PE geonets had quite low strength and could stretch 40%–50% while the PP

Figure 7.3 Pulling tests on a PP geonet side by side to the rutted old trail (to the right). The route have been passed by about 60 ATVs during work on a power dam upstreams. *(Photo I Wästerlund.)*

net had only 10% elongation and some PP nets could not even get to break but started to slide in the clamps (Type 13). Thus, we closed a PP net to be our main geonet to protect the ground, which proved to be good choice at the tests (Fig. 7.4).

Vehicle

Wheel inflation pressure was 40—50 kPa and the mass with 75 kg operator was between 420 (4WD) and 670 kg (6WD). The 4WD vehicle could produce traction of 3.5 kN. On the rutted swampy peatland, the maximum traction sank to 3 kN and on the destroyed trail the traction sank to 2 kN. The friction coefficient could be calculated to 0.65. On the geonets, the friction was 0.6, mostly because the lugs in the tire did not fit into the mesh to get proper grip.

Calculated as net traction (rolling resistance—max traction), it becomes rather small loads that can be pulled on a horizontal but destroyed alpine road (cf. Fig. 7.1). Practically, that means a vehicle pulling a trailer starts to slip and then breaks the ground, thus, the rolling resistance becomes heavier than the net traction and the ATV is stuck!

When similar pulling tests were made on a geonet, the net pulling force became almost double stronger (Fig. 7.5).

In summary for the vehicles:

- Suzuki a 13-year-old machine with worn out tires but with good balance and only 23 hp. Not powerful but ok in most cases at bad routes.

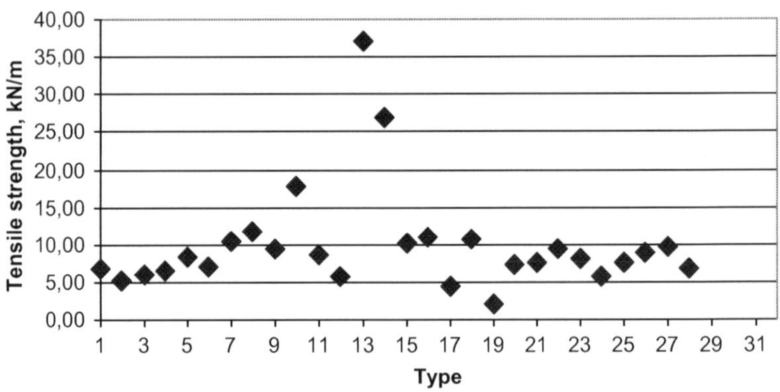

Figure 7.4 Test results from the tensile testing of varying types of geonets. Type nr 13 and 14 was a PP plane stiff geonet, that we used during further tests.

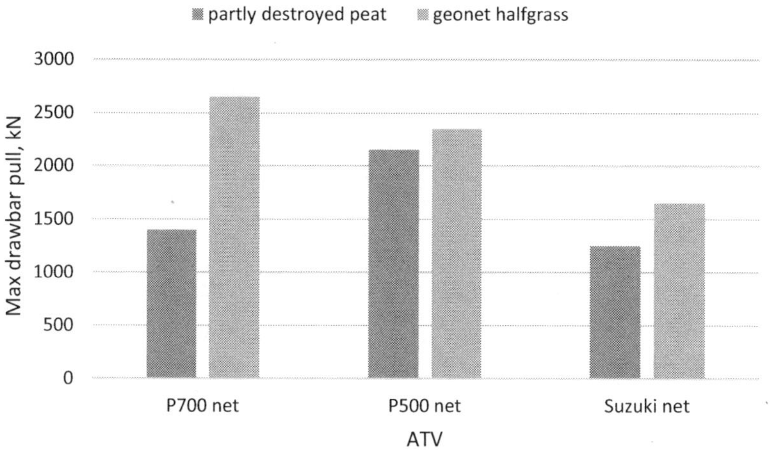

Figure 7.5 Measured maximal ATV net traction on different grounds with and without geonets and on ground with only Carex (sedge) and peaty ground.

- Polaris 700 a new powerful machine (45 hp) with new tires, but the big engine is not any benefit at half destroyed routes.
- Polaris 500. A new 6WD with a bogey rubber tracks over the rear wheels (40 hp) used for transport of equipment to the mountain areas.

The properties of the ground

The plate tests showed that the Carex meadow with some water running on top of the ground had a rather thick grass overlay the peatland and with a root layer 3–8 cm thick with 350 kPa strength. Below that mass of roots the strength sank to 200–250 kPa (Fig. 7.6). When the grass/root layer is destroyed, the strength is very weak and an ATV sinks down 5–8 cm in the ground which means the rolling resistance increases considerably. With a geonet on top of the ground, we increase the ground strength considerably. The shoe in Fig. 7.6 sinks down 2–3 cm, thus you will need a rather thick sole on the shoe to not get wet.

The strength of the geonets had to be measured with a hydraulic cylinder to be able to produce enough of force 500 kPa (trailer loaded with an ATV plus two persons to load the trailer). It showed that a PE net sank down into the peaty ground at only 450 kPa load. With anchored PP net, it could withstand 650 kPa until the peaty ground started to give away. Thus, with a pp net muck higher forces could be withstand, but still there is a great variability on the alpine peaty meadows. A PE net would increase the

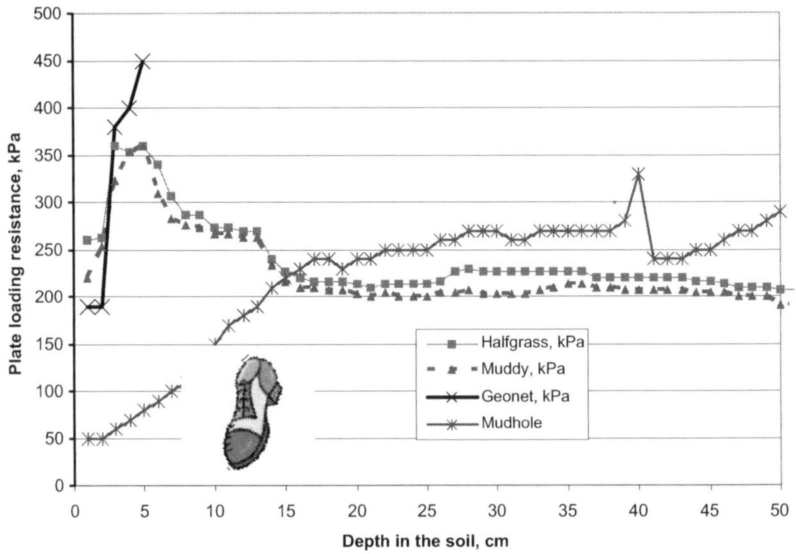

Figure 7.6 Measured plate strength with 10 cm^2 round plate on undisturbed half grass meadow (n = 3) and straight (X) on a half grass (sedge) meadow with geonet. Maximum applied force was 450 kPa. The shoe indicates a human foot pressure sinkage on the muddy ground.

strength considerably, but after 280 kPa it could be pressed down into the peaty ground. With a PP net, the load could be increased up 500 kPa, thus, double the strength of the peat bog.

Anchoring of the nets and vegetation

We learned two important things during our work on the alpine meadows. First thing was that the vegetation had to be, cut before putting out the net on the ground, because when the grass is bent down, it will choke the vegetation below it and their roots, which will weaken the mat. After cutting the new growth through the geonet could amount to 180 g fresh weight m^{-2} after 2 months of growth compared to 270 g m^{-2} without any net, thus although cutting of the vegetation and the net only somewhat reduced growth of the vegetation 2.5 months afterward. We had some proofs (only with a picture) showing that the geonet completely disappeared after 2 years but the net was still protecting the ground, not shown here.

The other thing was that the anchoring of the net must survive both the winter and hold the net on place. The first material we tested was the Norvia types of screws. It was a plastic type with fiber filling. But as Kiguchi (2005) noted in his report something is happening between the plastic material and the wood fibers and the whole thing becomes white and brittle. On top of that we discovered that the thawing during the winter lifted up the screws and the year afterward many were all lifted up 1–3 cm above the geonet (Figs. 7.7 and 7.8).

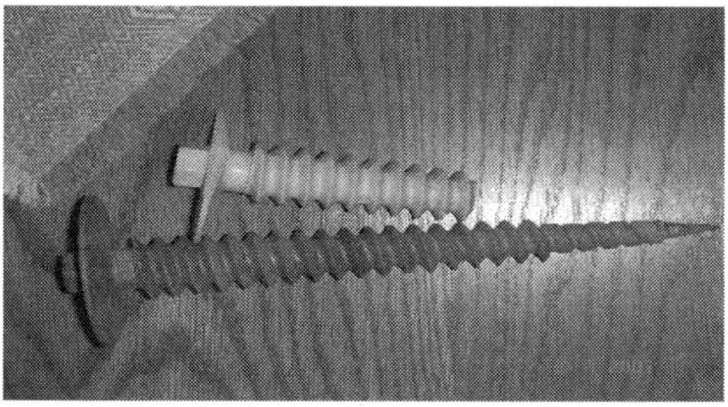

Figure 7.7 Examples of the Norvia screws, originally 22 cm long, one new with wood fiber filling (brown (gray in print version)) and one after 1 year after a winter outdoor, gray-white and broken.

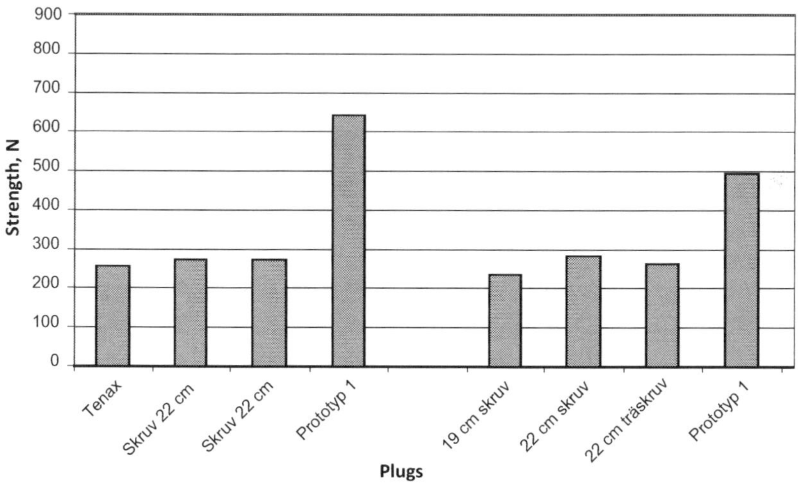

Figure 7.8 Pulling tests at 22 degrees angle of different types of anchoring and the anchors had been in the ground for 2 h before testing, during two seasons.

Thus, we had to make our own type. In our tests, we also realized that below 12–15 length there were of no use to have the anchor attached to peat mud. Mårten took his band saw and played around and made a prototype, which we called prototype 1. This one was with two legs and supplied with a bow to be able just to tramp it down and a shank for the two legs (Fig. 7.9).

In our tests it became almost completely outstanding compared to all other types including Tenax and Tensar types for anchoring of tents (Fig. 7.9). The benefit of having two legs was that we could get a hold on two or more meshes. Thus, we had now a whole system with both how to cut the vegetation and how to put out the geonet by rolling it out from a trailer after the 6WD ATV and anchoring + wooden bridges to overcome small streams not shown here.

This project was closed down without any further action from the county ministers involved in the project and the Sami people had to work further with any support.

Environment

The first criterion was the reindeers. Most of them had no problems to walk on the geonet. We learned also from the Sami people that a reindeer calf had hoofs that demand a mesh size 30 mm × 30 mm to be able to walk on the geonet without getting stuck.

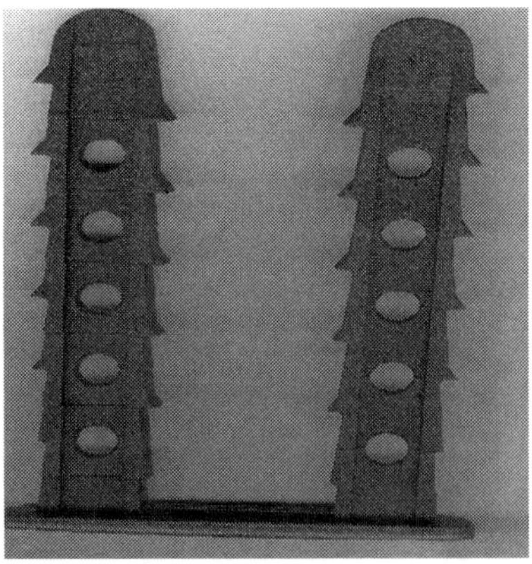

Figure 7.9 The factorial design of prototype 1 used to anchor the geonets and one per meter net was enough. *(Photo: I.Wästerlund.)*

The other aspect is if something can leach out from the plastics. But it is a completely inert material. Testing done at Luleå Technical University (Lindebergh and Kappfjell, 2006) showed that the material was totally inert and had no leaching substances.

Third aspect, if you regret the net, it is very possible to just pull up the net. If carefully pulled up, there will be no residues at all left on the ground compared to wooden bridges often used to reinforce walking path! (Fig. 7.10).

Practical recommendations

1. The geonet must have a strength of at least 20 kN m^{-1} to be able to hold for an ATV (twice times safety)
2. The net should be plane stiff, thus, very little elongation in the longitudinal direction and not stretch that cuts the vegetation or bend down in holes.
3. The net should have some warts to give some grip for the ATVs tires.
4. The hypothesis that the net reinforce the ground may not hold: It is more likely to protect the vegetation and the systems good strength (net + anchoring + vegetation + roots).

Figure 7.10 Our project leader surprised after been driving back and forth 10 times on a peat bog with PP net anchored with what we called prototype 1 and he could just return back to road without any marks after a 2 tonnes car. He stands beside his car just surprised.

5. The anchoring should be at least 400 N at 20 degrees pulling angle, but the length of anchoring could be only 15–20 cm long to the depths.
6. Ditches should NOT be dug in the peat bogs but put in drainage tubes that will be just tramped down and the net just put over it to let water run to the other side.
7. It is an environmentally friendly system and the mesh size should be 25 mm × 30 mm to fit for reindeer calves not getting stuck in the net, no harmful substances, and can be pulled up when not in use without leaving any external material.

Final test

We let our project leader take his car and ran 10 times back and forth on our newly reinforced peat bog. He just stepped out from his car and looked surprised. It was a 2 tonnes car with 4WD and there were no marks after the driving (Fig. 7.10)!

DIXI.

References

Allard, A., Löfgren, P., Sundquist, S., 2004. Damages on ground and vegetation in three alpine areas due to summer driving. SLU, Dept of Forest Resorce Management and Geomatics, Umeå. Report No126.

Anon, 1999. TENAX Gound Reinforcement. Wrexham, UK. Available at: www.Tenax uk/r/index. Php?page=trm.

Kiguchi, M., 2005. Weathering performance of wood-fibre plastcs. Presentation at XXII UIFRO World Conference 8 – 13 August. Brisbane, Australia.

Lindbergh, J., Kappfjell, M., 2007. Markförstärkningsmattor i fjällmiljö: En miljöanalys av plast- och stålnät. ITU. Luleå Technical University.

Sirois, D.L., Stokes, B.J., Rawlins, C.L., 1989. Cone penetrometers –how do they measure up? ASAE/CSAE Paper No89-7067.

Wästerlund, I., 2006. Environmentally friendly alpine roads with help of geonet reinforcement. In: .Proc. 10th International ISTVS Conference. 3–6 Oct 2006, Budapest, Hungary.

Whitlow, R., 1990. Basic soil Mechanics, second ed. Longman Scientific & Technical, Essex.

Wong, J.Y., 1978. Theory of ground vehicles. John Wiley & Sons, New York, Chichester, Brisbane, Toronto.

CHAPTER 8

Logging residues as soil–root protection*

What do we mean by logging residues?

Nordic forestry is to major part mechanized, where the trees are felled by a single-grip **harvester** which has a harvester head in the tip of the 5–6 m long crane and feeding rollers feeding the tree trough the knifes to delimb the tree (Fig. 8.1). With a chain-saw in the head the tree is felled but also to cut it into suitable lengths, suitable for the factory buying the tree (sawmill or pulp factory or other types of factories). This done at the place where the tree was standing, but delimbing can be done be done at a separate places to give a proper pile of logging residues.

The forwarder collects the stems from the ground with 6–12 m long crane with a grip that can be both tiltable and rotating and place them on the loading area. Each stem piece could weigh somewhere between 200 and 800 kg. With 50 stem pieces the load could be 10,000 to 40,000 kg on a 10–20 Mg vehicle. Driving in 5 km h^{-1}–20 km h^{-1} in the terrain, 200–2000 m with the load in the terrain.

Figure 8.1 A pile of logging residues after the harvester after delimbing spruce trees.

* Major part of Effect of logging residual reinforcement of strip roads on rutting and soil compaction on a moist fine-grained soil section was first published in Forest and Ecology Management 252 (2007) 118–123.

Soil and Root Damage in Forestry
ISBN 978-0-12-822070-2
https://doi.org/10.1016/B978-0-12-822070-2.00008-0
© 2020 Elsevier Inc.
All rights reserved.

The stems are transported to road side with a forwarder, which carries the load from the terrain to the road and unloading at the landing, to be collected and transported to a factory by a timber lorry, carrying may be 40–70 tonnes on the road.

Logging residues consist of the foliage of trees and tree tops (branches and crowns) unmarketable bolts and undergrowth trees. They form the most notable and also economically significant source of raw material in the production of wood fuels and constitutes roughly about 20% of the bioenergy supply to Swedish energy supply out of 40% of the total energy supply (in Sweden). There is, however, a great variation in the amount and composition of logging residues between different felling sites. Tree species, standing volume, size, the branches of trees, and the extent of decay influence the quality of logging residues (Savolainen and Berggren, 2000). Roughly a Norway spruce tree (*Picea abies* (L) H. Karst.) can have more than 50% foliage compared to trunk mass, whereas pine trees (*Pinus sylvestris* L) might have 20%–25% foliage compared to trunk mass and deciduous trees may have 15%–20% branches.

All trees have tree tops with a rather high bark proportion, often around 20%–50% according to Nurmi, 1993. Roughly, a tree contains 40% cellulose, 30% hemicellulosa, and 20%–30% lignin, and about 40% of a tree is the stem (Bodig and Jayne, 1982).

Tree species

In the Nordic countries, not only softwoods are most common like Norway spruce (*Picea abies* (L.) H. Karst.) and Scots pine (*Pinus sylvestris* L.) but also hardwoods like aspen (*Populus tremula* L.) and Silver birch (*Betula pendula* Roth, formerly *Verrucosa* Ehrh.) are very common.

The amounts of logging residues depend on the age of the stand and tree species. A cleaning stand (young trees—spacing) could have anything between 20 m^3 and 150 tonnes ha^{-1} according to the National Forest Surveys in Sweden, and according to own analyses of data.

In many countries, the material is measured by volume and in others by weight (and control of MC, Moisture Content). The drawback is that the material is naturally quite wet and usually about 50% moisture content (MC dry) of the weight. On the other hand, most of the material consists of branches from conifer trees. Branches are supported by compression wood (to hold up the needles and twigs on conifer trees) (see Fig. 8.2). Conifer tree branches are lifted up, and deciduous trees have the support on the

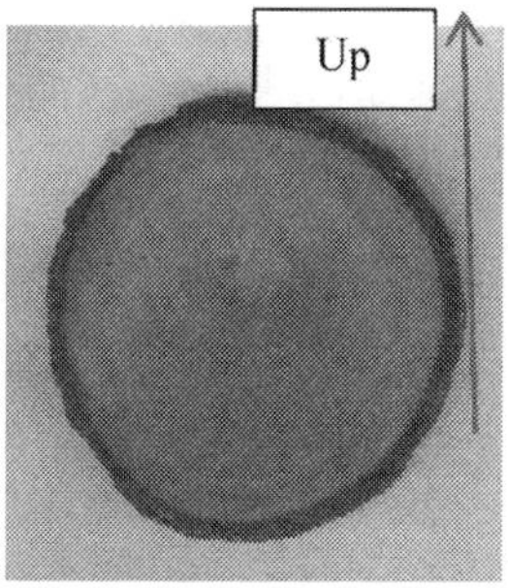

Figure 8.2 Cross section of a thick spruce branch (diameter 55 mm) showing the heavy reinforcement on the underside to lift up the branch (own material).

upper side to hold up the branches up (tension wood); both the tension and compression wood contain high levels of lignin and both have high energy content (20–35 kJ kg^{-1} DM (dry matter)) due to gluing with lignin of wood cells (Fig. 8.2).

According to Nurmi (1993), in both conifer and deciduous trees (except *Populus tremula*), which have much lower content, the lignin content is 22%–30%, having an effective heating value of 19–20 kJ kg^{-1} DM (Nurmi, 1993).

Components in the material

The juvenile wood in tree tops contains mostly cellulose and hemicellulose (called holocellulose together). This is a young type of wood formation usually occurring in tree tops containing less of lignin but bark may be more occurring and may be more than 20% and bark has a rather high effective heating value (19–21 kJ kg^{-1} DM according to Nurmi, 1993). Chipped together with the branch wood, the lignin-rich branch wood will constitute the major part of the weight and thus give a rather high energy content (around 19–20 MJ kg^{-1} for conifers) giving a decent energy content for the crown. The major problem with logging residues is high amount of volatiles 0.5%–2% and ash 0.5%–4% depending on the amount of bark.

Branch wood has a higher calorific heat value compared to stem wood (18–20), and it increases with decreasing branch diameter (Nurmi, 1993). The calorific value is the inherent energy in the material usually determined on laboratories (usually with a bomb calorimeter). After that determination adjustments are made for hydrogen (H) in the water produced during combustion, roughly 6% according to Nurmi (1999). This figure is called effective heat value (W_{eff}). The resulting figure is denoted as **at delivery** and this figure is then adjusted for actual moisture content in the fuel. Energy is measured either in kJ kg^{-1} or as kWh kg, the latter being a common measure for district heating centrals. Water in the material could be either given as moisture content on wet basis (MC wet) or moisture content on dried basis (MC dry), the difference being drastic at wet levels (Fig. 8.3).

Effect of logging residual reinforcement of strip roads on rutting and soil compaction on a moist fine-grained soil

Introduction

In Swedish forestry, harvesting operations are usually fully mechanized, i.e., carried out by a harvester and a forwarder. In final felling operations (i.e., clear-cutting), there is a risk for rutting and soil compaction, which can be considerable under unfavorable soil conditions, as the machines used are heavy (Fig. 8.4). A large harvester can exceed 20 Mg and a laden large forwarder approaches 40 Mg. Soil disturbance has mainly been studied in thinning operations where small- to medium-sized machines have been used (Wronski and Murphy, 1994; Jansson and Wästerlund, 1999). Only a few studies have been made where medium to large forwarders (20–38 Mg) have been used for wood transport. These studies' reports increase in rut depths with increasing number of machine passages (Wronski et al., 1990; Jansson and Johansson, 1998; Eliasson, 2005) and of soil

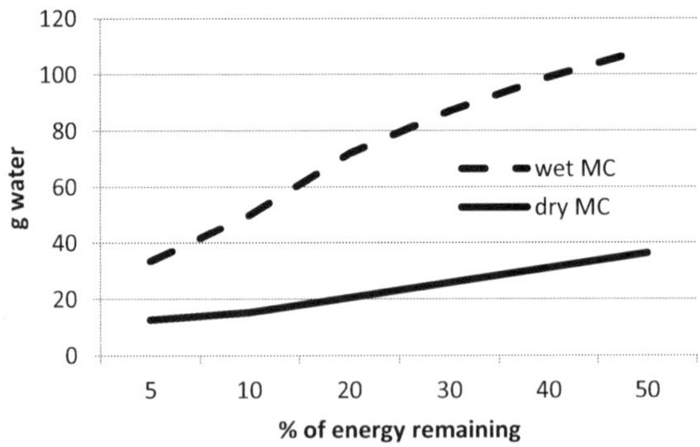

Figure 8.3 Energy content versus moisture content with comparison between wet weight as a base or dry weight (MC wet vs. MC dry).

Figure 8.4 The loaded forwarder passing our trail. *(Photo Wästerlund.)*

compaction reaching deeper than 40 cm (Jacobsen and Greacen, 1985; Jansson and Johansson, 1998; Eliasson, 2005).

In a review of the effects of soil compaction on forest crops, Wronski and Murphy (1994) note that a major disadvantage with machine traffic in thinning is that roots of living trees are damaged which cause growth losses in the forest crop. Tree roots are an important strength component in forest soils increasing soil strength considerably (Björkhem et al., 1975; Wästerlund, 1989), thus improving trafficability in the forest stand. In final felling, this effect can be used without risk for growth losses due to root damage as the tree crop is harvested. However, if the new forest crop is established in compacted soil, there is a risk for reduced growth (Sands and Bowen, 1978; Wästerlund, 1985; Wronski and Murphy, 1994) and reduced value of the wood when harvested (Murphy et al., 2004).

To avoid rutting and soil compaction, logging debris (slash), i.e., branches and tree tops, is used to reinforce strip roads in the stand during harvesting operations (Wronski and Murphy, 1994). Studies have shown that slash reinforcements of strip roads reduce both rutting (Hallonborg, 1982; Wronski et al., 1990) and soil compaction (McDonald and Seixas, 1997) and result in a substantial increase in apparent soil-bearing capacity (Wronski et al., 1990). However, there are rather few studies of the effects of slash reinforcements on soil density changes after traffic with large forwarders on moist fine-grained soils.

During the last decade, there has been an increased interest in the Nordic countries to utilize logging debris for energy production. To get an efficient handling of the material and to avoid contamination of the fuel with soil particles, the logging debris is concentrated in piles on the side of the strip roads. Thus, no reinforcement strip roads are done except where it is absolutely necessary from a trafficability point of view. The extraction of the residues and in some cases bundling or chipping of the residues increase the machine traffic in the strip roads and, thus, the risk for rutting and soil compaction. A further complication is that the most interesting stands for extraction of logging residues are spruce stands as they have a high stocking and, thus, a high concentration of logging residues per ha. Spruce is probably the tree species in Sweden that is suffering most when growing in compacted soils (Wästerlund, 1985). The most fertile spruce sites are situated on moist and fine-grained soils which have low bearing capacity, and there is thus a high risk for ground damage and negative effects on future tree growth. This has increased the need for a deeper understanding of where and in what quantities slash is needed for strip road reinforcement.

Aim

The aim of the study was to (1) evaluate if slash reinforcement of strip roads influences rutting and soil compaction caused by traffic with heavy forest machines on a moist fine-grained soil and (2) quantify these effects of slash reinforcement of strip roads.

Materials and methods

The study was done in one Norway spruce (*Picea abies*) stand close to Nyåker in northern Sweden (63.8460490N, 19.8160230E). The stand was growing on a silty clay sediment (Fig. 8.5) situated at the base of a slope. The study was designed as a factorial experiment with two factors, slash depth in the strip road, and number of machine passages, replicated in four blocks. The effects of 0, 10, and 20 cm slash in the strip roads (approximately 50% and 100% of the slash available within the normal working width of the harvester) and one, two, and five machine passages were studied.

The first passage was made with the harvester, a 20.5 Mg Timberjack 1270B, during harvesting of the stand. As work progressed, the harvester reinforced the strip road in front of the machine with logging debris, branches, and tops, according to the experimental design. The tires on the front bogies were Nokian ELS 600/55-26.5 and on the rear axle Nokian TRS L-2600/65-34. Tire loads on the front bogies were 3.1 Mg per tire

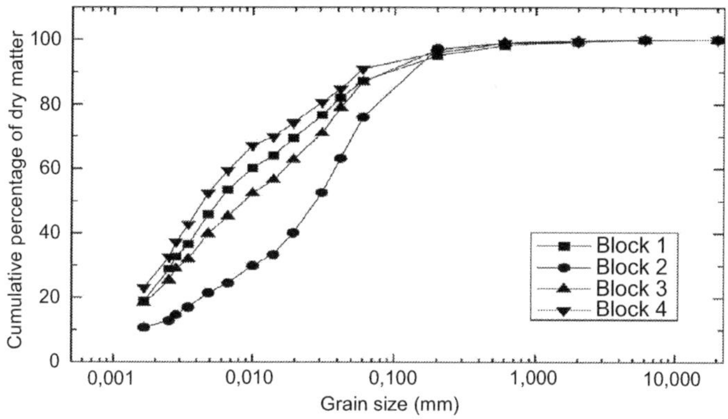

Figure 8.5 Particle size distribution of the soils in blocks 1–4.

and on the rear axle 3.9 Mg per tire. All wheels were filled with approximately 70% water, and the bogies were fitted with tracks and the rear wheels with chains. The second to fifth passage were made by a laden large forwarder, Timberjack 1710, totaling 37 Mg including the load. The same load of pulpwood was used throughout the whole experiment. Tire loads on the forwarder front part were on average 3.5 Mg per tire and on the rear part 5.75 Mg per tire. On the forwarder, both front and rear bogies had Nokian ELS 750/55-26.5 tires and were fitted with tracks, Olofsfors ECO-TRACK, due to track use all tires had an inflation pressure of 600 kPa.

Blocks were established to minimize variations in soil structure and soil moisture within blocks (Fig. 8.5, Table 8.1). In each block, three 20 m long plots were established and measurement of soil density, soil water content, and cone resistance was made before trafficking and after one, two, and five machine passages. Rut depths were measured after one, two, and five machine passages. Measurements were made in six permanent sample points per plot, i.e., in each track at 7, 10, and 13 m distance from the start of the plot, except for soil density and soil water content, which were measured in four sample points per plot, i.e., in each track at 7 and 10 five passages on four permanent sample points per plot except for soil density and soil water content, which were measured in four sample points per plot, i.e., in each track at 7 and 10 m distance from the plot start. The strata gauge was placed perpendicular to the direction of the rut and placed to be as centered as possible in the expected rut. The two holes for the gauges probes were filled with wooden rods between measurements, enabling measurements in the exact same location during the whole experiment,

Table 8.1 Initial soil dry density (g cm^3), soil water content (g cm^3), and air-filled pore volume (cm^3 cm^3) in the four blocks.

	Block			
Depth (cm)	**1**	**2**	**3**	**4**
Dry density (g cm^3)				
10	0.772^a	0.436^b	0.762^a	0.657^a
20	0.942^{ab}	0.853^a	1.080^b	0.925^{ab}
30	1.150^a	1.123^a	1.209^a	1.133^a
40	1.325^a	1.208^b	1.360^a	1.333^a
Water content (g cm^3)				
10	0.374^a	0.404^b	0.399^{bc}	0.383^{ac}
20	0.424^a	0.457^b	0.444^{ab}	0.429^a
30	0.447^a	0.471^b	0.454^a	0.447^a
40	0.446^a	0.464^b	0.453^{ab}	0.448^a
Air-filled pore volume (cm^3 cm^3)				
10	0.351^{ac}	0.424^b	0.304^c	0.367^{ab}
20	0.217^a	0.207^a	0.134^a	0.219^a
30	0.116^a	0.087^a	0.074^a	0.122^a
40	0.050^{ab}	0.061^a	0.016^b	0.045^{ab}

Values on the same row followed by different letters were significantly separated ($P < .05$).

m distance from the plot start. Slash depths were measured between the ruts at 8.5 and 11.5 m distance from the plot start, both prior to and after traffic, and all slash in a 0.3 m square were collected in these spots after traffic. These slash samples were separated in three fractions based on the diameter of the material (1 < 10 mm, 10 mm < 25 mm, and 125 mm) and the green mass of each fraction was measured (Table 8.2). Soil samples were collected, using 50 mm high steel cylinders with 72 mm diameter, from 20 to 40 cm

Table 8.2 Slash thickness and slash density separated on slash thickness class.

	Slash thickness class		
	0	**10**	**20**
Slash thickness (cm)	0	9.9	20.4
Slash density (kg m^2)			
Ø<10 mm	0	5.3	12.8
10 mm ≤ Ø < 125 mm	0	1.7	4.5
Ø ≥ 25 mm	0	0.6	0.4
All fractions	0	7.6	17.7

depth adjacent to the middle sample point in the left track of each plot prior to traffic. After determination of water content (105°C), these samples were used for sieving analyses and determination of plastic and liquid limit as well as particle density. Average soil particle density was 2.58 g cm^3.

Rut depths were measured as the distance between the underside of a wooden rod laid across the rut and undisturbed ground surface, the measurement was made centrally in the rut. Slash depth was measured as the distance between the ground surface and the top of the slash bed when the slash bed was compressed with hand force.

Soil wet density and soil water content were measured with a Campbell Pacific nuclear dual probe strata gauge MC-S-24 at 10, 20, 30, and 40 cm depth in the track after zero, one, two, and as soil water readings from the strata gauge are known to have a risk for systematic deviations from the true values (Jansson, 1998). Soil water content readings were controlled using a comparison between relative water content from the soil samples and the relative water content from the same depth with the strata gauge. Regression was used to get a function for correction of the water content measured by the strata gauge. After correction of all water contents measured, soil dry density was calculated as the difference between wet density and water content. Air-filled pore volume (V_a) was calculated as

$$V_a = 1 - V_w - \frac{Dd}{\rho}$$

where V_w is the water content, *Dd* the soil dry density, and ρ is the soil particle density.

Pretraffic dry density, water content, and air-filled pore volume differed between blocks (Table 8.1) but not between treatments. Pretraffic soil cone resistance between 10 and 20 cm depth in blocks 1–4 were 0.99, 0.52, 0.69, and 0.88 MPa, respectively. The humus layer was thicker in block 2 (14 cm) than in blocks 1, 3, and 4 (7, 9, and 6 cm, respectively). The field layer was dominated by bilberry (*Vaccinium myrtillus* L.) in blocks 1, 3, and 4, and by sphagnum mosses and bilberry in block 2.

Plot mean values were used for all measurements in the analyses, thus n = 48 for each soil depth or when changes were analyzed n = 36. For comparisons of the effects of slash depth, differences from soil measurements prior to traffic were analyzed using analysis of variance in SPSS (release 11.0.0).

Treatment effects were considered significant if $P < .05$.

Results

Rut depths increased with increased number of machine passages ($P = .027$) (Table 8.3) but were not influenced by slash depth ($P = .171$) or the interaction between number of machine passages and slash depth ($P = .764$). The ANOVA model had a low adjusted R^2, 0.131. Analyses were not improved when slash density was used as a covariate instead of the factor slash depth.

Slash reinforcement of the strip roads significantly reduced the changes in soil dry density and soil porosity in the upper 20 cm of the soil (Tables 8.4 and 8.5), although no mixing of slash and soil was observed. No effects of slash reinforcement could be found on soil water content (Tables 8.4 and 8.7). An increased number of machine passages increased soil dry density in the upper 20 cm and reduced the air-filled pore volume in the upper 30 cm of the soil profile (Tables 8.4 and 8.6). Machine traffic wetted the soil in the top 10 cm, and there was a tendency that soil water content increased with the number of passages, below 10-cm depth soil became dryer as the number of machine passages increased (Tables 8.4 and 8.6). In the uppermost 20 cm of the soil, observed soil dry density increase after one machine passage on strip roads without slash was higher than after five machine passages where strip roads were covered with 20 cm of slash (Fig. 8.6).

At 10 and 20 cm depth, the relative increase in soil dry density could be estimated with the following functions:

$$\frac{\rho_{diff}}{\rho_0} = 0.281 + 0.105\ln(N_{pass}) - 0.129d_{slash}$$

at 10-cm depth and

$$\frac{\rho_{diff}}{\rho_0} = 0.130 + 0.040\ln(N_{pass}) - 0.045d_{slash}$$

Table 8.3 Rut depths separated on slash depth class and number of vehicle passages.

	Slash thickness class (cm)		
No. of passages	**0**	**10**	**20**
1	5.2	4.1	4.4
2	5.5	4.8	4.6
3	9.3	5.7	6.7

Table 8.4 Significance levels (*P*-values) for the effects on change in soil dry density, soil water content, and air-filled pore volume, respectively, by block, slash depth, passage, and the interaction of slash depth and passage from the ANOVAs for each depth.

Depth	P-Value Block	Slash	Passage	Slash passage
Dry density (g cm³)				
10	0.091	0.000	0.001	0.845
20	0.003	0.000	0.015	0.935
30	0.000	0.807	0.100	0.970
40	0.002	0.380	0.270	0.926
Water content (g cm³)				
10	0.006	0.220	0.052	0.859
20	0.000	0.131	0.017	0.865
30	0.007	0.360	0.003	0.815
40	0.026	0.289	0.001	0.661
Air-filled pore volume (cm³ cm³)				
10	0.064	0.001	0.002	0.929
20	0.000	0.001	0.014	0.894
30	0.000	0.987	0.058	0.965
40	0.001	0.465	0.308	0.939

At 20-cm depth, where ρ_{diff} is the dry density increase, ρ_0 the initial dry density, N_{pass} the number of passages, and d_{slash} is the slash depth in dm. Adjusted r^2 was 0.501 at 10-cm depth and 0.366 at 20-cm depth and at both depth all variables were highly significant ($P < .01$).

Changes in soil density were significantly ($P < .001$) negatively correlated with changes in air-filled pore volume at all depths (r = 0.984 to 0.919). At 10 cm depth, soil density changes were positively correlated with the changes in soil water content ($P < .001$, r = 0.556), but below 10 cm depth no correlation between the two could be found. Rut depths were significantly ($P < .05$) correlated with changes in soil density at 10 and 20 cm depth (r = 0.411 and 0.333), but not with soil density changes at 30 and 40 cm depth.

Discussion

Forest soils in the Nordic countries are usually heterogeneous and soil properties may vary within small distances. To minimize the influences of

Table 8.5 Slash thickness in the strip road versus changes in soil dry density (g cm^3), soil water content (g cm^3), and air-filled pore volume (cm^3 cm^3), respectively.

	Slash thickness class		
Depth (cm)	**0**	**10**	**20**
Dry density (g cm^3)			
10	0.212^a	0.135^b	0.074^b
20	0.159^a	0.101^b	0.066^b
30	0.075^a	0.064^a	0.065^a
40	0.069^a	0.057^a	0.043^a
Water content (g cm^3)			
10	0.012^a	0.016^a	0.010^a
20	0.010^a	0.005^a	0.007^a
30	0.022^a	0.018^a	0.020^a
40	0.022^a	0.025^a	0.021^a
Air-filled pore volume (cm^3 cm^3)			
10	0.093^a	0.068^{ab}	0.039^b
20	0.051^a	0.034^{ab}	0.019^b
30	0.007^{aa}	0.007^{aa}	0.006^{aa}
40	0.004^{aa}	0.003^{aa}	0.004^{aa}

Values on the same row followed by different letters were significantly separated ($P < .05$).
Values followed by "aa" are not significantly separated from 0.

soil heterogeneity when studying impacts of machine traffic, measurements preferably should be made with nondestructive methods in the same spot. When tracking changes in soil density, a dual probe gamma gauge may be one of few tools for nondestructive measurements (cf. Rawitz et al., 1982; Jansson, 1998), as the measurements can be made repeatedly in exactly the same holes for the measurement probes if these are filled with wooden rods between measurements. The wooden rods had no effect on comparisons between treatments or on the soil compaction process, although they did absorb some moisture during the experiment.

Slash thickness is difficult to measure accurately, both depending on what pressure to apply before measurement and due to variations of the slash thickness within a strip road, and thus a large number of measurements are needed to get a good estimate. Furthermore, the effect of slash thickness on rutting and soil compaction depends on slash properties (size, strength, etc.). The mass of slash and slash size distribution within a defined area could give a better estimate on these effects but is more difficult to apply for practical use.

Table 8.6 No. of machine passages versus changes in soil dry density (g cm^3), soil water content (g cm^3), and air-filled pore volume (cm^3 cm^3), respectively.

	No. of passages		
Depth (cm)	**1**	**2**	**3**
Dry density (g cm^3)			
10	0.093^a	0.129^a	0.198^b
20	0.080^a	0.106^{ab}	0.140^b
30	0.055^a	0.058^a	0.092^a
40	0.043^a	0.053^a	0.073^a
Water content (g cm^3)			
10	0.010^a	0.011^a	0.017^a
20	0.004^a	0.012^b	0.006^{ab}
30	0.015^a	0.025^b	0.019^{ab}
40	0.016^a	0.028^b	0.024^b
Air-filled pore volume (cm^3 cm^3)			
10	0.046^a	0.061^a	0.094^b
20	0.027^a	0.029^a	0.048^b
30	0.006^{aa}	0.003^{aa}	0.016^a
40	0.000^{aa}	0.007^{aa}	0.004^{aa}

Values on the same row followed by different letters were significantly separated ($P < .05$).
Values followed by "aa" are not significantly separated from 0.

Rut depths increased when the number of passages increased but no treatment difference was found. However, in general, ruts were shallow except for in block 2, where a thicker humus layer in combination with a wetter soil led to deeper ruts. Furthermore, in block 2, the humus largely consisted of sphagnum moss and peat in contrast to other blocks where it mostly consisted of decomposing litter and roots.

In studies with similar objectives, it has been observed that slash cover on strip roads provides soil protection (Jacobsen and Moore, 1981; Wronski et al., 1990; McDonald and Seixas, 1997), the results of the present study confirm this. The thin slash cover (10 cm) gave less protection than the thick especially after multiple passages. With increasing number of passages, wear and breakage of the slash cover probably contributed to reduce the effects of slash, especially when the slash cover was thin (Wronski et al., 1990); however, no significant interaction between slash thickness class and number of passages was found. It is probable that the type of slash, i.e., depending on tree species and size of trees harvested, will influence the

Table 8.7 Amount of slash and time used for the preparation.

Treatment	Repetition	Time (sec)	Amount (tonnes)	Thickness (cm)
Slash	1	624	6.3	22
	2	619	6.7	25
Geonet	1	1597	2.9	17
	2	828	2.2	10
Logs	1	?	Average diameter 13 cm	

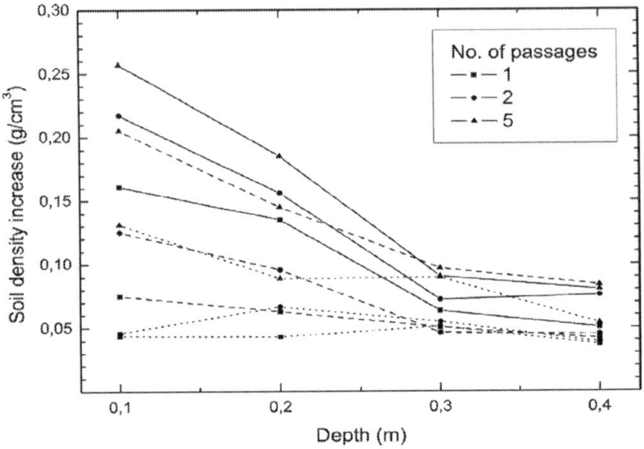

Figure 8.6 Observed soil dry density increase (g cm^3) versus slash reinforcement treatments (no slash = *solid line*, 10 cm *slash hatched line*, and 20 cm *slash dotted line*) and number of passages (one passage = *squares*, two passages = *circles* and five passages *triangles*).

effect of slash on soil compaction and the depth to which slash influences soil compaction. Furthermore, soil type and soil moisture content will affect the effects of slash on soil compaction. These relationships need further studies.

In the present study, the use of slash from a spruce forest as a cover on the strip road reduced relative soil compaction at 10 cm soil depth by 12.9% per 10 cm of slash cover and at 20 cm soil depth by 4.5% per 10 cm of slash cover. Thus, the major effects of slash seem to occur in the top soil, which also is the most important for spruce tree growth (Wronski and Murphy, 1994). When using the functions predicting relative dry density increase care must be taken not to use them outside of the material they are based on. The functions are based on slash depths between 0 and 2 dm and logically avery thick slash bed might prevent soil compaction but it will not cause a reduction in soil density, as long as no mixing of slash and soil occurs.

The absence of significant treatment differences in soil dry density changes at 30 and 40 cm depth can partly be explained by low potential for compaction, due to a combination of almost water-filled pores and a fine soil that restricted soil water movements.

In order to reduce the risk for soil disturbance without using logging debris to reinforce the strip roads technical solutions such as larger tires, lower tire pressure, and bogie tracks can be used, i.e., methods similar to those used on agricultural vehicles (Alakukku et al., 2003; Chamen et al., 2003). An increase of the tire width from 600 to 800 mm reduced rut depths with approximately 50% for an eight-wheeled 22 Mg forwarder (Myhrman, 1990). As there are legal limits on machine width on public roads and the tire of large forwarders usually are between 700 and 750 mm wide, the possibility to further increase tire width is limited. Due to high wheel loads and uneven terrain, with obstacles such as stones and stumps, tire pressures on large forwarders are held high compared to on agricultural machinery. Reducing tire pressure on a laden large forwarder from 600 to 300 kPa tended to reduce soil compaction in the uppermost 20 cm of the soil during the first forwarder passage; however, after four forwarder passages, no differences in soil compaction could be found (Eliasson, 2005). As a large share of the soil compaction were caused by the harvester and as it is difficult to reduce harvester tire pressure without compromising machine stability and thus the ability to reach out with the boom, other solutions are needed to reduce soil disturbance caused by the harvester. One solution could be to use bogie tracks, which was found to reduce rut depth and cause comparable changes in cone index after traffic when used on a trailer with a bogie axle load of 12.5 Mg compared to when the trailer was used without bogie tracks (Bygden et al., 2004). Due to the risk of track inflicted damage to the tires, tire manufacturers recommend that tire inflation pressures were kept close to maximum levels when bogie tracks are used. In the present study, the rear bogie of the loaded forwarder carried 62% of the total load. An even mass balance would reduce the rear axle load with approximately 4.5 Mg, which could reduce the soil compaction to some extent (Haåkansson and Petelkau, 1994).

To reinforce strip roads with the amount of slash used in the present study approximately 24 Mg slash per ha are necessary for the 10 cm treatment and 56 Mg slash/ha for the 20 cm treatment, assuming 12.5 m between strip roads and a strip road width of 4 m. There are further benefits of leaving the slash in the cut than just a reduction of soil compaction, as it is shown that removal of the slash has negative effects on the growth of the new stand (Egnell and Valinger, 2003). On the other hand, the income from slash sales to heating plants contributes to the net revenue of the cut and is said to partly pay for site preparation and planting of the site at the same time as it makes site preparation and planting easier. Site preparation has an amelioration effect on compacted soil (Miwa et al., 2004), but not all of the compacted soil will be affected by site preparation and the amelioration effects are dependent on the site preparation technique used.

If heating plants would accept forest fuel contaminated by soil particles, slash could be used both to reinforce strip roads and thereafter as forest fuel. However, at present, this is not the case as it would necessitate improved chipping and combustion techniques.

This study shows that soil compaction in the top soil, so important to growth, can be reduced by slash reinforcements of the strip roads. It would seem worthwhile to reinforce strip roads with slash on sensible parts of clear-cuts even when slash is harvested for energy use. Further studies are needed to evaluate if it is possible to further reduce soil disturbance by proper combinations of slash reinforcement and bogie tracks or by a combination of slash reinforcement and reduced tire pressures.

Studies are also needed to clarify how much of the compacted soil that is loosened by scarification and to which depth scarification loosens the soil. Studies are also needed to clarify how much of the compacted soil that is loosened by scarification and to which depth scarification loosens the soil.

Reducing ground damage during slash forwarding

Introduction

The Swedish energy policy to increase the share of renewable sources.

The goals pointed out that the energy supply should be safe, environmentally friendly, plus be at reasonable prices. Bioenergy is an important part of that and logging residuals (slash). Wood fuel is the most used energy source and during the last twodecades it has increased with 73% (Anon, 2008) (Fig. 8.7).

Figure 8.7 The Forwarder with full load of slash.

With increased collection of wood fuel from the Swedish forests, the transports are also increasing, and therefore the loads on the soils and according to Froehlich (1979): soil compaction. Soil damage and rutting can occur.

The most productive soils are spruce stands on moist, and thanks to high moisture content in the ground clear-cutting, it leads to high ground water level and therefore vulnerable for soil damage with use of heavy machines. Logging residues on the ground may reduce soil damage, but the best protection is from branches more than 10 mm in diameter (Eliasson and Wästerlund, 2007).

Aim

The aim is to evaluate if there are alternative to logging residuals to minimize rutting and soil compaction and to study rutting on unprotected soil. Aspects like working time to handle the material and the damages.

Material and methods

The field experiment was done on a 10-ha clear-cut area in the middle of Sweden. It had been cut as a winter cutting 1 year earlier, and in September it was the ordinary logging residuals forwarding for collection of the material.

During the experiment, a John Deere 1510e (forwarder) from 2008 adapted for slash forwarding with extra banks and a ducktail (elongation of the rear end) to be able to take bigger volume of slash since slash is rather light. The 6WD had Nokia Forest King F tires 710/45-26.5 on all wheels and the rear wheels were in a bogie and supplied with ECO-OF tracks for traction. The empty forwarder with tracks and banks weigthed 18,440 kg and during the experiment the forwarder was loaded with 10 tonnes of logging residues, thus totally 28,440 kg. Ground pressure calculated according the RTG formulas (Malmberg, part 2, 1981) was 75.4 kPa for the front wheels and 56.3 kPa for the rear wheels with track.

Four different treatments were tested. (1) unprotected, (2) slash bedded cross-over the travel direction, (3) thin stems, 20–10 cm in diameter, and 1 and (4th) a geonet with a thin layer of logging residues to protect the geonet from the studs in the tracks.

According to Byggros, the geonet we used was Tensar RN 2530, a plane stiff net (30 kN m^{-1}) used for stabilizing purpose produced from propen with square pattern with 39 mm × 39 mm openings and it was anchored with RN brackets.

Well out on the area, it turned out to be possible to put out only seven trails and thus the repetitions had to be cut down. They were put out as similar soil conditions as possible and marked with sticks at each corner. The bed of slash was done by the operator of the forwarder with the slash put cross the track.

The geonet was rolled out by hand and anchored at every second meter with the brackets. Stumps were cut as close to the ground as possible. Unprotected ground was cleaned from logging residues as clean as possible by hand. After these operations, every fifth meter was marked to give three penetrations per marked spot with Eijkelkamp penetrometer on both sides of the track. The tracks were 20–25 m long.

Time to put out the layer of slash on the ground started when the crane started to move and stopped when it was folded back. The slash load was weighted before and after with help of the built in hydraulic cylinders in the ALS system.

When all preparations were done, the driving began and time to pass the track with all wheels was measured. Driving speed the operator could chose himself considering both conditions and the risk for the load to tip.

On unprotected tracks, the rutting depths were measured from a rod put over the wheel-rut and the depth measured from the rod to the deepest part of the rut after 2,4, and 6 passings excluding stones or branches.

This measurement was done on four places after the geonet was lifted away with the crane. After the last driving both rut depth and cone resistance were measured.

The cone resistance in the ground was measured on 15b points before driving and at 6 points after the driving. Later the data were emptied over to a tap top and examined in Excel.

On all trails, 25 m was measured up and with a height meter for trees the inclination was measured and all trails were in the direction of the slope which were between 4 and 7 degrees. Soil samples were taken on all trails on 10 and 20 cm depth (Fig. 8.6) and drying showed 55% on the upper 10 cm and below 9%–10%.

The cost for putting out the geonet and thin layer of slash was calculated according to the following:

Cost per m = (consumption of slash × the cost for the slash) + (use of geonet and brackets) + (cost for geonet and brackets) + (time consumption and time cost for forwarder)/(length of trail).

The cost for putting out the slash layer was calculated according to the following:

The cost per meter = (use of slash × cost for slash) + (time for forwarder × costs for forwarder)/(length of trail).

Calculation model:

Cost per meter = (time cost for the forwarder cost × time)/(length of trail).

Amount of slash during the experiment was noted in tonnes and the price for the forwarder was 900 SEK (10 SEK = 1 EURO), the cost for slash was 300 SEK per tonnes, and the cost for geonet was 35 SEK m^{-1} and the TN brackets 15 SEK per piece.

Results

Table 8.6 shows the time it took to put out the different types of soil protection and the layer thickness at the different trails:

Logs

Already during the first passage some logs started to crack and at next passage it broke (softwood), which caused the rutting depth to increase to 9 cm after 6 passages. The average soil compaction is shown in Figs. 8.8 and 8.10.

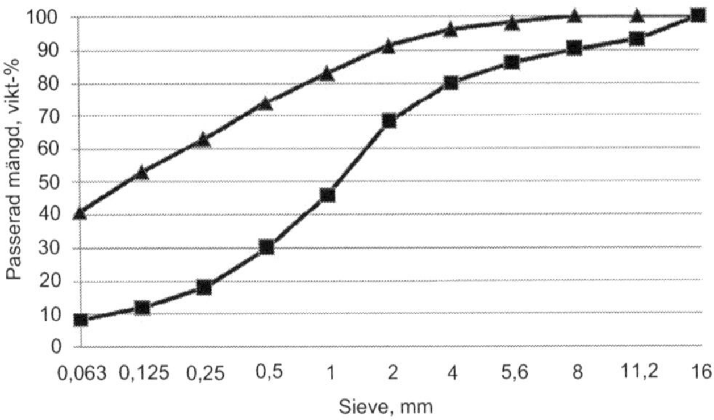

Figure 8.8 Particle size for soil samples distribution from ca.20 cm depth.

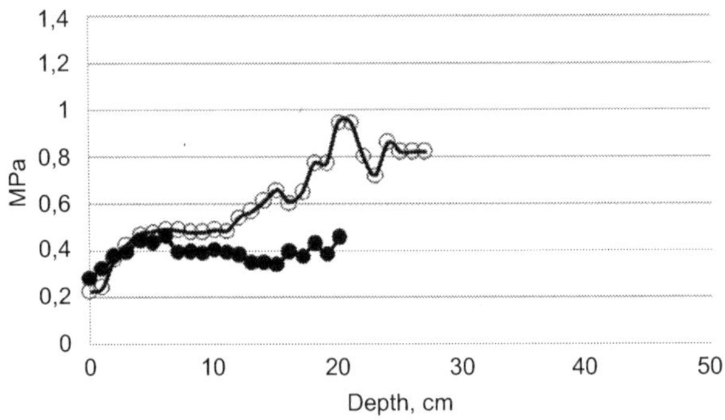

Figure 8.9 Cone resistance in the ground before passage (*unfilled rings*) and after six (*filled rings*) passages with a bed of pulpwood.

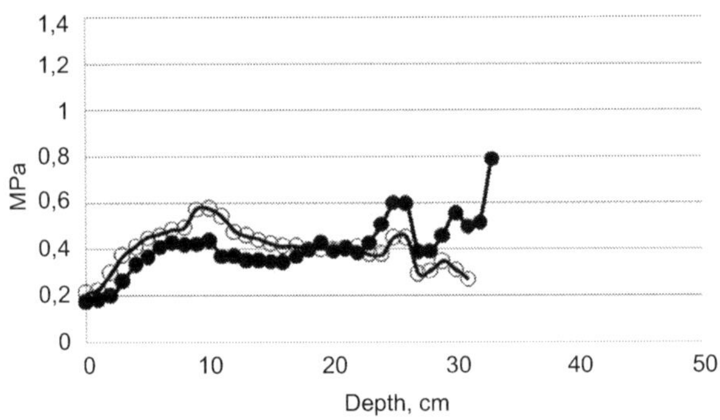

Figure 8.10 Cone resistance in the ground before passage and after six passages on thick slash layer. *Unfilled rings* after 6 passages on a 22 cm slash layer. Every point is an average of 15 observations. Trail 1.

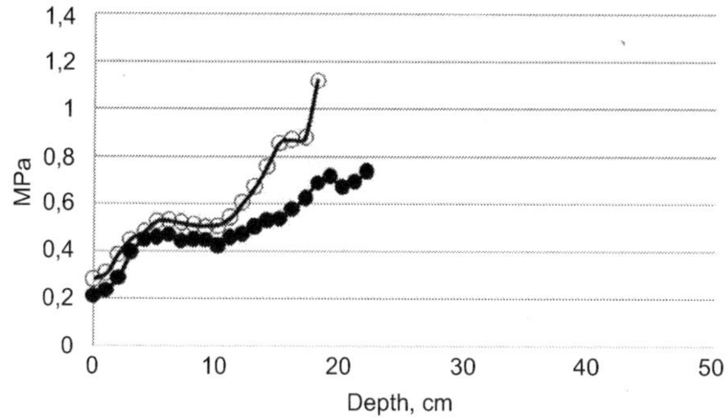

Figure 8.11 Cone resistance in the ground before passage (*unfilled rings*), average of 15 observations) and after six passages (*filled rings*) on a 25 cm thick slash layer thick slash layer. Trail 2.

Slash

On both places with a slash layer a 20 cm, it was possible to see a hint of rutting only at the last passage with the forwarder. The average rutting was after first passage 5 cm and after second passage 7 cm. The soil compaction is shown Figs. 8.11 and 8.12. Some loosening occurred above 20 cm.

Geonet

The first passage on the geonet with a thin layer of slash and looked good till the third passage when the geonet broke on a stump that lifted up the geonet. At the same problem occurred the forth but now on several places but the geonet managed to protect the ground and passage and the soil

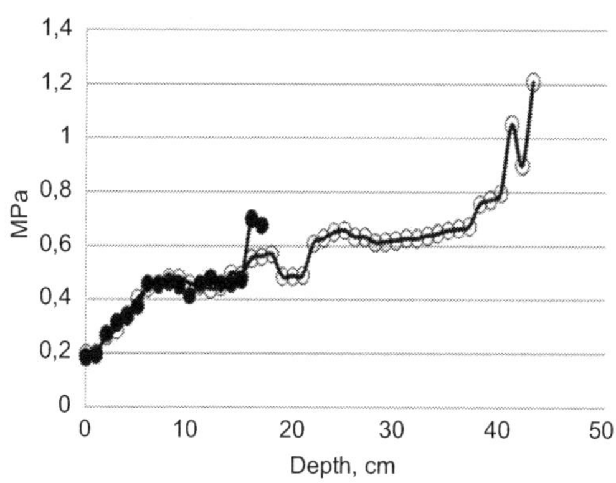

Figure 8.12 Cone resistance in the ground before passage (*unfilled rings*) and after six passages (*filled rings*) on slash and geonet.

showed no rutting. Great parts of the geonet was intact and under the geonet. The average rutting depth became only 6 cm and no change in soil compaction as shown in Fig. 8.9 after 6 passings. At the second trail, damages occurred already at second passing, where a point load caused the geonet to crack and at following passing's the geonet broke totally and some RN brackets broke. Average rutting became 19 cm and some loosening of the soil showed that the soil was influenced (Figs. 8.11, 8.12 and 8.13).

Unprotected (control)

Already during the first passage on unprotected soil, rutting occurred in the soil. At every additional further deepening occurred. After 6 passages, there was 37 cm deep rutting and on the second trail there was 50 cm deep rutting. A clear influence on the soil could be measured after passing 1 and 2. In the upper soil, some loosening occurred to become soil compaction in the depth as shown in Figs. 8.14 and 8.12.

In Fig. 8.15, a compilation of cone resistance is given for the major trails including their repetition. Standard deviation was 0.09 and 0.23 before any passing per point and depth (n = 105 per cm) in the interval 1−15 cm. After 6 passings, the standard deviation was 0.09 and 0.28 in the interval 1−15 cm.

But a clear change occurred after 15 cm depth where a clear soil compaction is visible.

A compilation over the rutting development on the unprotected trails per passing is shown in Fig. 8.16. Fig. 8.18 is an average from the eight

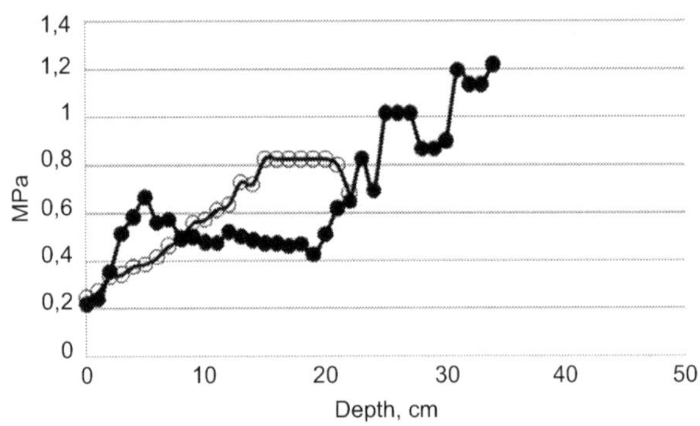

Figure 8.13 Cone resistance in the ground before passage (*unfilled rings*) and after six passages on slash and geonet 1 (*filled rings*) with 10 cm slash. Every point is an average of 15 observations.

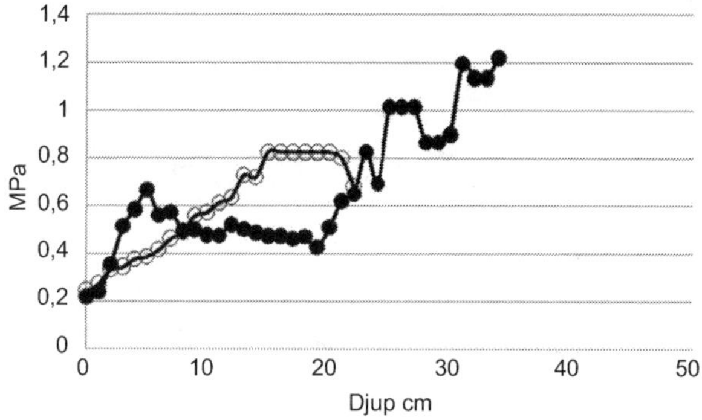

Figure 8.14 Cone resistance in the ground before passage (*unfilled rings*) and after six passages on slash and geonet 2 (*filled rings*) with 10 cm slash. Every point is an average of 15 observations.

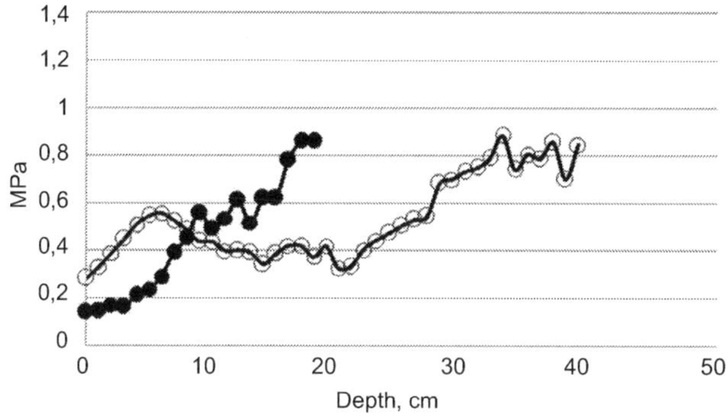

Figure 8.15 Cone resistance in the control ground before passage (*unfilled rings*) and after six passages (*filled rings*) on unprotected ground.

measurement points and passing no. 2, 4, and 6. An analysis of variances in Minitab showed that it can be described as a line to 97.5% between passing 2–6.

Table 8.8 shows a compilation of the time measurements done during the experiments. There were some problems. Two values are pointing up. The first value depends on the forwarder got stuck in the mud and the second case was a stump where it got stuck. In both cases, it was on unprotected soil (Control).

The cost for the forwarder is the measured time for the forwarder times to put out the soil protection. The cost for the geonet is based on how

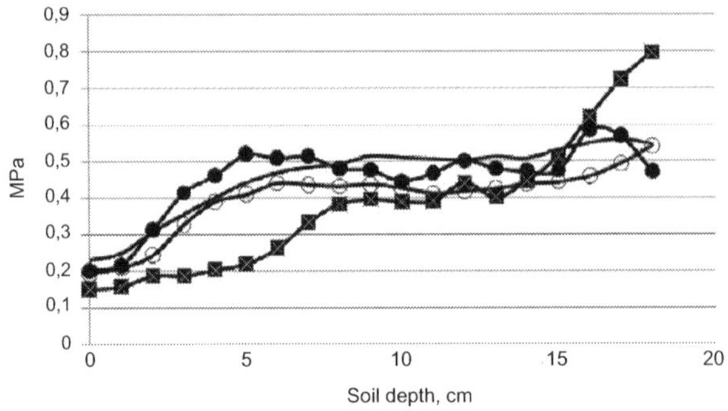

Figure 8.16 Cone resistance in the ground before passage and after six passages on plots with repetition. *Line* with

■ is the unprotected (Control),

● plots with geonet and

○ plots with slash and *line* without any dots are before passing.

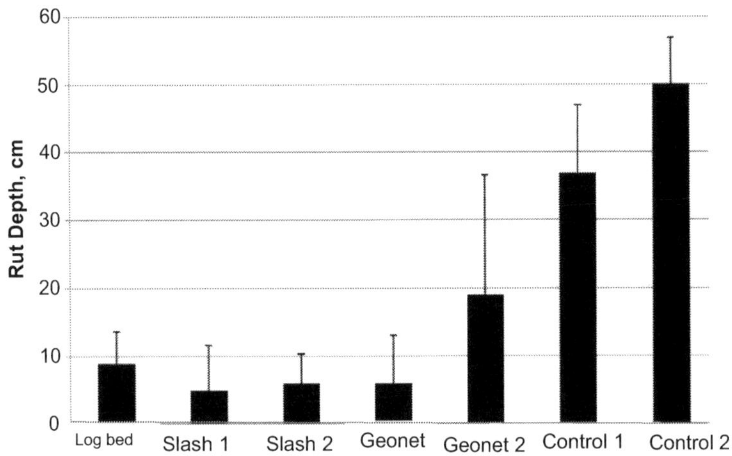

Figure 8.17 Average wheel-rut for all plots after six passages. Bar indicate max–min.

much geonet that was used and the amount of RN brackets used per 100 m length.

The cost for slash is calculated based on the amount of slash used per trail and then multiplied to 100 m. On the last row, the total cost for soil protection means are summarized. The biggest cost is for the geonet and also slash is a great post (Tables 8.9 and 8.10).

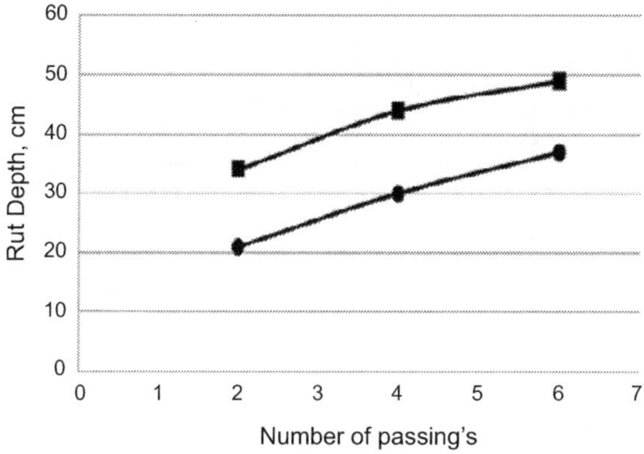

Figure 8.18 Wheel-rut development with increasing number of passages on not protected plots.

Table 8.8 Amount of time at crossing the plots.

Treatment	Repetition	Passing 1	2	3	4	5	6
Log bed		64	48	44	45	38	46
Slash	1	49	44	33	33	31	33
	2	71	49	51	46	48	45
Geonet	1	55	50	45	49	54	50
	2	59	44	43	44	44	40
Control	1	42	38	61	56	70	100
	2	127	61	62	58	84	58

Discussion

To be able to do any statistical analysis was not possible due to few repetitions. The use of a cone penetrometer may not be optimal. There are a number of factors that will influence the reading but the soil in the forest is seldom homogenous. The cone penetrometer to document the changes in the soil may not be optimal, but on the other hand there are very few alternatives that work in such environment, and it has been used in a number of cases to study soil changes after vehicle passing.

Price for the slash may vary between different places and companies. The weight may have an uncertainty but it affects quite little (± 300 kg), influence quite little the total sum. The loaded forwarder may be pressed in some water in the top soil and therefore some loosening was noted but in deeper soils compaction was noted.

Table 8.9 Cost calculation for sec/100 m ground cover with geonet + a thin layer of slash.

	Slash		Geonet	
Cost type	**Trail 1**	**Trial 2**	**Trial 1**	**Trial 2**
Machine	624	558	1397	833
Geonet	0	0	15,725	15,725
Slash	7414	7885	3413	2589
Totally	8038	8443	20,535	19,147

Table 8.10 Average cost calculation for crossing the different 100 m trials.

Trial	Log bed	Geonet	Slash	Control
Kr/100 m	46	47	42	62

In Table 8.4, it is shown the pattern how the soil protection influences the time for driving and the cost for the forwarder. However, it looks to be quite big difference but means quite little for slash forwarding.

For the geonet to compete with slash, the price for geonets must be cheaper. The geonet should not cost more than 7 sek m^{-2} to be able to compete with slash and then delete the RN brackets ska kunna konkurera med GROT som markskydd måste inköpspriset bli betydligt lägre. According to these data and without RN brackets, the geonet may cost 11 SEK m^{-1} to be compatible with slash.

At the moment, there are a geonet Tensars Triax with a tensile strength 47 kN m^{-1} (Minsér, pers medd 2010), costing 22 SEK m^{-2}, since it is made with a new method with punching out with triangles, which mean that the total cost will sink to 10,500 SEK per 100 m.

An interesting aspect is how much slash one can get for the geonet per 100 m according to Fig. 8.19. It is about 53.5 tonnes GROT and calculated to a thickness it would become 48 cm. If we compare it with the cheaper geonet (22 kr m^{-2}) with slash, the breaking point will be 36 tonnes and the slash thickness 32 cm.

In summary: all tested soil protection mean here worked but to very different costs. But as far as no costs have to be paid for soil damage. On the other hand, wheels may press down organic material into the top soil and loosening the soil and ameliorate the soil compaction. But in deeper soil, there might be increased compaction.

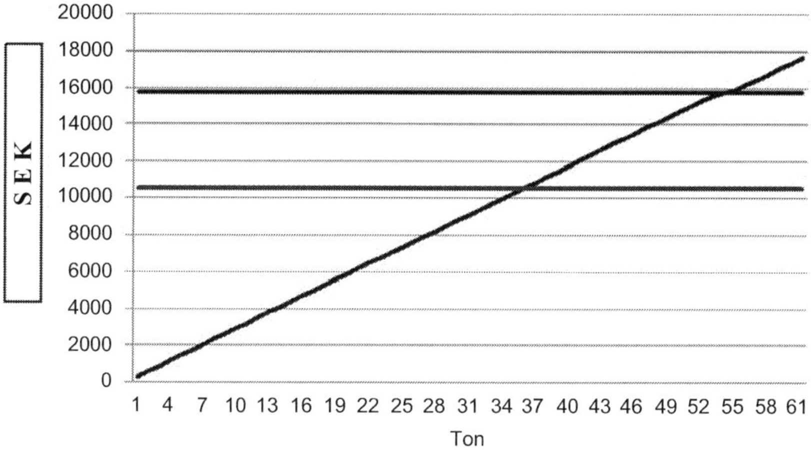

Figure 8.19 Breakpoint where the costs for 100 m geonet 1 with anchors, 100 m geonet 2 with anchors are equal to the amount of slash (10 SEK = 1 EURO).

Conclusions

- There are alternatives to use SLASH to reduce soil damage but to relatively high costs,
- Geonet did not fit really to the uneven clear-cutting area but may be it will be god for areas with small patches with peatland,
- As long as there are no fines for soil damage, nobody will put out soil protection.

References

Alakukku, L., Weisskopf, P., Chamen, W.C.T., Tijink, F.G.J., van der Linden, J.P., Pires, S., Sommer, C., Spoor, G., 2003. Prevention strategies for field traffic-induced subsoil compaction: a review. Part 1. Machine/soil interactions. Soil Tillage Res. 73 (1/2), 145–160.

Anon, 2008. Energy Indicator 2008. Available at: www.byggros.com; info@byggros.com.

Björkhem, U., Lundeberg, G., Scholander, J., 1975. Root Distribution and Compressive Strength in Forest Soils. Root Mapping and Plate Loading Tests in Thinning-Stage Stands of Norway Spruce. Departments of Forest Ecology and Forest Soils, Royal College of Forestry, Research Notes, Stockholm, 60 pp.

Bodig, J., Jayne, B., 1982. Mechanics of Wood and Wood Composites. Van Nostrand Reinhold Company, New York.

Bygden, G., Eliasson, L., Wästerlund, I., 2004. Rut depth, soil compaction and rolling resistance when using bogie tracks. J. Terramech. 40 (4), 179–190.

Chamen, T., Alakukku, L., Pires, S., Sommer, C., Spoor, G., Tijink, F., Weisskopf, P., 2003. Prevention strategies for field traffic-induced subsoil compaction: a review. Part 2. Equipment and field practices. Soil Tillage Res. 73 (1–2), 161–174.

Egnell, G., Valinger, E., 2003. Survival, growth, and growth allocation of planted Scots pine trees after different levels of biomass removal in clearfelling. For. Ecol. Manag. 177 (1−3), 65−74.

Eliasson, L., 2005. Effects of forwarder tyre pressure on rut formation and soil compaction. Silva Fenn. 39 (4), 549−557.

Eliasson, L., Wästerlund, I., 2007. Effects of slash reinforcement of strip roads on rutting and soil compaction. For. Ecol. Manag. 248, 118−123.

Froehlich, H.A., 1979. Stickvägar och körskador i gallringsbestånd (Soil compaction from logging equipment: Feffects on growth of young ponderosa pine). J. Soil Water Conserv. 34, 276−278.

Haåkansson, I., Petelkau, H., 1994. Benefits of limited axle load. In: Soane, B.D., van Ouwerkerk, C. (Eds.), Soil Compaction in Crop Production. Elsevier, Amsterdam, pp. 479−499.

Hallonborg, U., 1982. Ristäckets Inverkan På Spårbildningen. Resultat 3. Forskningsstiftelsen Skogsarbeten, Stockholm, 4 pp. ISSN 0280-1884.

Jacobsen, B.F., Greacen, E.L., 1985. Compaction of sandy forest soils by forwarder operations. Soil Tillage Res. 5, 55−70.

Jacobsen, B.F., Moore, G.A., 1981. Effects of two types of skidders and of slash cover on soil compaction by logging of Mountain ash. Aust. For. Res. 11, 247−255.

Jansson, K.-J., 1998. Effects of machinery traffic in forestry on soil properties and tree growth. PhD ihesis. In: Faculty of Forestry, Swedish University of Agricultural Sciences, vol. 66. Acta Univ. Agric. Suec. Silv., Uppsala, 24.

Jansson, K.-J., Johansson, J., 1998. Soil changes after traffic with a tracked anda wheeled forest machine: a case study on a silt loam in Sweden. Forestry 71 (1), 57−66.

Jansson, J., Wästerlund, I., 1999. The effect of traffic by light forest machinery on the growth of young Norway spruce trees. Scand. J. For. Res. 14 (6), 581−588.

Malmberg, C.-E., 1981. Terrängmaskinen del:2 [Off Road Vehicle, part 2. Forest Research Institute Stockholm.

McDonald, T.P., Seixas, F., 1997. Effect of slash on forwarder soil compaction. J. For. Eng. 8 (2), 15−26.

Miwa, M., Aust, W.M., Burger, J.A., Patterson, S.C., Carter, E.A., 2004. Wetweather timber harvesting and site preparation effects on coastal plain sites: a review. South. J. Appl. For. 28 (3), 137−151.

Murphy, G., Firth, J.G., Skinner, M.F., 2004. Long-term impacts of forest harvesting related soil disturbance on log product yields and economic potential in a New Zealand forest. Silva Fenn. 38 (3), 279−289.

Myhrman, D., 1990. Factors influencing rut formation from forestry machines. In: Proceedings of the 10th International Conference of the ISTVS. Kobe, Japan.

Nurmi, J., 1993. Heating values of whole-tree biomass in young forests in Finland. Acta Forestalia Fennica 236, Tampere, Finland.

Nurmi, J., 1999. The storage of logging residue for fuel. Biomass Bioenergy 17, 41−47. Fennica. 236. Tampere, Finland.

Rawitz, E., Etkin, H., Hazan, A., 1982. Calibration and field testing of a two probe gamma gauge. Soil Soc Am. J. 46, 461−465.

Sands, R., Bowen, G.D., 1978. Compaction of sandy soils in radiata pine forests. II. Effects of compaction on root configuration and growth of radiata pine seedlings. Aust. For. Res. 8, 163−170.

Savolainen, V., Berggren, H., 2000. Wood fuels basic information pack. Jyväskylä and Sollefteå. Benet.

Wästerlund, I., 1985. Compaction of till soils and growth tests with Norway spruce and Scots pine. For. Ecol. Manag. 11, 171−189.

Wästerlund, I., 1989. Strength components in the forest floor restricting maximum tolerable machine forces. J. Terramech. 26 (2), 177–182.

Wronski, E.B., Murphy, G., 1994. Responses of forest crops to soil compaction. In: Soane, B.D., van Ouwerkerk, C. (Eds.), Soil Compaction in Crop Production. Elsevier, Amsterdam, pp. 317–342.

Wronski, E.B., Stodart, D.M., Humphreys, N., 1990. Trafficability assessment as an aid to planning logging operations. Appita 43 (1), 18–22.

Further reading

Anon, 1980. Riksskogstaxerings Inventering Åren 1978-79. Swedish National Forest Inventories 1978-79 SLU, department of Operational Efficiency, Garpenberg. Report no 137.

Eriksson, L., 1981. Stickvägar och körskador i gallringsbestånd (Striproads and Damages Caused by Machines when Thinning Stands). SLU, Department of Operational Efficiency. Report No 137. (in Swedish, summary in English).

Index

Note: "Page numbers followed by "t" indicate tables, "f" indicate figures and "b" indicate boxes".

U

V

W

Made in the USA
Monee, IL
13 September 2026

63138401R00111